Caprock Canyonlands

NUMBER EIGHTEEN

The M. K. Brown Range Life Series

Big Butte in the Tule Canyon Basin, an immense fin of red Dockum sandstone

Journeys into the Heart of the Southern Plains

Caprock Canyonlands

BY DAN FLORES

with photographs by the author

University of Texas Press

First edition, 1990
Printed in Hong Kong

Portions of certain chapters of this book have been published in different form as "Forgotten Canyonscapes of the Plains," Persimmon Hill *16 (Summer 1988) and "Canyons of the Imagination,"* Southwest Art *18 (March 1989).*

♾ The paper used in this publication meets the minimum requirements of American National Standard for Information Sciences —Permanence of Paper for Printed Library Materials, ANSI Z39.48-1984.

Excerpt from the poem "Mountain Lion," in The Complete Poems of D. H. Lawrence, *edited by Vivian de Sola Pinto and F. Warren Roberts (Penguin Books, 1977). Reprinted by permission of Viking Penguin Inc., New York.*

LIBRARY OF CONGRESS CATALOGING-IN-PUBLICATION DATA

Flores, Dan L. (Dan Louie), 1948–
Caprock canyonlands : journeys into the heart of the southern plains / Dan Flores. —1st ed.
p. cm. — (M. K. Brown range life series ; no. 18)
Includes bibliographical references.
ISBN 0-292-71121-2 (alk. paper)
1. Canyons—Southwest, New. 2. Southwest, New—Description and travel—1981– 3. Landscape protection—Southwest, New. 4. Flores, Dan L. (Dan Louie), 1948– . I. Title. II. Series.
F787.F48 1990 89-36616
917.904'33—dc20 CIP

TITLE PAGE:
Blanco Canyon at dawn

Slot canyon in
Los Lingos narrows

To the late
Georgia O'Keeffe,
who loved the
Southern Plains
and captured
their magic,
and to
Alexandre Hogue,
who spoke out
against their
destruction.

Contents

Gypsum layers in the Permian Formation, South Cita Canyon

I will tell you a pleasant tale which has in it a touch of pathos.

—Mark Twain,
Letters from the Earth

Midway in a four-day hike through Palo Duro and Tule canyons

Preface

HE OLD-TIME NEW MEXICANS had a saying: "Hay las sierras debajo de los llanos"—There are mountains below the plains. Modern travelers crossing the Southern Plains on the interstates from Oklahoma City to Albuquerque or San Antonio to Santa Fe might doubt it, but the New Mexicans were right. Below the level of the flat horizon, great canyons carve mesas and buttes, spires and badlands through the architecture of the Llanos of West Texas and New Mexico.

This plains canyonland country, although not drawn on a scale comparable to the now-famous gorges and canyons of farther west—the Grand Canyon, Yosemite, the Snake River Gorge, Zion and Bryce and Canyon de Chelly and all those other stupendous canyons of the Colorado Plateau—was in technologically simpler times well known across the Southwest. The canyons were oases of water and trees and wildlife in the limitless expanse of the High Plains, the *sanctum sanctorum* of 120 centuries of Native American occupation. You've seen portrayals of them, although not in the flesh since Monument Valley or California usually stunt doubled for them, in dozens of Westerns. Tucked away in the recesses of the plains, they're mostly forgotten now except by landscape artists, victims of an odd quirk resulting from privatization on the western plains, the tendency (especially marked in Texans) to speak in terms of counties and ranches rather than landforms and rivers.

But they are still here, the old names clinging to them out of a time that recedes from our view like embers winking out: Largo, Tule, Blanco, Los Lingos, Palo Duro. And of a sudden, newly distinctive, recognized as an ecological subregion in the Southwest, and as the one subregion on the Southern Plains where the best expression of the wild diversity of the original plains yet exists.

On the face of it, there are good reasons why I should not be writing a book about the Llanos canyonlands. I am not a native of the Southern Plains, in the sense of being born here, but am instead a "down-streamer." For two centuries, and on one side of my family line for nearly three, my ancestors have been woodland folk from the lower Red River. My mother's family, Hales and Temples from the Anglo-Celtic border class, almost springs from the ground of the western Arkansas hills. My dad's, Zylkses and Lafittes and Floreses from the long-ago ethnic mix of the Louisiana-Texas frontier, has been associated with a Louisiana Red River tributary called Bayou Pierre since the eighteenth century. Which causes me to wonder if some cumulative environmental gene imprinting might not be at work on me. Shouldn't my blood reflect, just slightly, the chemical content of the waters of Bayou Pierre, or the minerals of Petit Jean Mountain in the Ozarks? If so, why is it that I find the arid, sunlit canyons so intensely compelling? Brought up to the moist and the verdant, I am in love with the sere and the minimal. Did some long-ago ancestor, perhaps my great-great-grandfather, Pierre Flores, the product of converging lines of Indian traders and Red River pioneers, decide to go see where the Red River came from? And actually trace the river into Palo Duro Canyon, imprinting my DNA with a recognition code that struck me, a century later, with staggering *puha,* as the Comanches would say?

I don't know. But for years I thought that something like this was the explanation for a recurring childhood dream of mine, of a landscape image like a Georgia O'Keeffe painting, no action or people, just a suspended dreamscape of a vermilion canyon wall under a cobalt sky, with white clouds like hanging cotton balls, pervading all of a sense of . . . Indians! In high school I decided that I must have seen, when very young, a dramatic postcard or a *National Geographic* photograph that had made a strangely profound impression. Later, in college, I resolved on the more metaphysical explanation that the dream was a genetic memory and anticipated the rush of recognition in my wanderings through the West.

At a Christmas gathering in the early 1980s I learned that at least part of the answer was not so metaphysical. In 1952 an aunt had moved to and lived for a short time in West Texas. As a four-year-old I had been taken along on a visit and somewhere, no doubt, had seen the dreamscape, a subliminal memory that I retained. But that did not explain to me why that particular image should have resonated through my dreams for so many years. So I still subscribe to the idea of ancestral imprinting and think that this is why I have always faced west, so to speak, and have had the sense of something very powerful whispering to me from that direction and no other.

I moved, in 1978, to West Texas and almost at once began to explore, with the weirdest sense of *déjà vu,* those canyonscapes of my childhood dream. And in 1983 I bought a small place in Yellow House Canyon, one of the headwater canyons of the Brazos River. Over five years, in a give-and-take as gradual and unceasing as the erosion of the overhanging rimrock, the canyon has worked its ancient magic on me, the kind of magic that has been reaching out of these canyonlands and wrapping around the human psyche for 12,000 years. This adult experience has combined with the child's fixation to become my motivation for writing this book, which is not, of course, the same thing as my goal in writing it.

I want to draw environmental attention to this country. If, as historian Alfred Runte and others have argued, the establishment of public parks and nature preserves is an expression of higher culture, on a par with the creation of great music, art, and literature, then as the rest of the Southwest has long suspected we are not guilty of having created much high culture on the Southern Plains. History, as I will endeavor to show, explains this neglect, and I won't dwell on the issue here except to say that there are compelling reasons, both ecological and societal ones, for a large-scale and imaginative alteration of the present situation. Some of those who stand for the status quo on the plains may find the philosophical basis of my arguments radical or offensive. That's a timeworn (and wornout) position on the plains. I will address it merely by saying that it is true that I am not especially impressed by the status quo in a landscape that I love. And to borrow Thoreau's defense, why anyway should I speak for the status quo? As he said, any churchman (or banker in Amarillo and Lubbock) would do that. I prefer to speak for the wild and for the sensuous heart.

Beyond that, this is a book of exploration and discovery, to a certain extent personal and experiential, of the human interaction with a landscape through time. I cannot lay claim, as Frederick W. Turner does in *Beyond Geography,* to writing "spiritual history." But this might be called a natural history of the spirit of a place in the American West.

DAN FLORES, *January 1989*
Yellow House Canyon, Texas

The prairie is undoubtedly the largest in the world, and the cañons are in perfect keeping with the size of the prairie.

—Explorer
George Wilkins Kendall,
1841

The high Llano in eastern New Mexico—a great sweep of horizontal yellow

Caprock Canyonlands

The change in the look of the land takes place suddenly and dramatically. . . . On top there is the plain of Texas, dryish but undramatic. Below, the red, eroded sandstone and the cactus . . . declare that this is New Mexico a good many miles before the mapmakers have recognized the fact.

—Joseph Wood Krutch,
The Desert Year

. . . the land is very different from what I had imagined.

—Explorer
Francisco Amangual,
1808

The dazzling badlands of the Little Red River and Caprock Canyons State Park

Chapter 1

Land of the Inverted Mountains

ON TOP—and it is not a syntaxic accident that we speak of being "on" plains as opposed to "in" woods, mountains, desert, or canyons—is a country possessing a sweep and expanse rivaled by few places on earth. To the upright human eye it appears utterly flat, the plane formed between earth and sky a perfect right angle. Lie on your back and the sky on a clear day is an overarching bowl of fathomless blue, a glittering, pulsing dance of lights on a clear night, the horizon line encircling the periphery of your sight as if you lie in a vast dish. Easy to see why humans have long been seduced by the meditative qualities of such a minimal landscape. Only the flat white of the polar regions has a similar power to drive the human mind back upon itself.

In recorded time the High Plain has borne many names, and before history was recorded on paper it must have had many more. The name that has stuck is Llanos, the Spanish word for plains, or Vegas, which carries a similar idea. In other parts of the world such country has been called *pampas, elmas, veldts, steppes,* all of them implying in their local dialects a landscape that is flat, high, windy, grassy, and arid or nearly so. Among the native peoples of the Southwest, some of whom lived here and most of whom traveled across it, names bearing the meaning "horizontal yellow" were common, a choice of double adjectives that, as is often the case with Indian names, catches the essence better than any European noun.

The Llanos have always had recognizable boundaries, the whole forming a giant oval lying east of the Sangre de Cristo range of the southern Rockies. The true Llanos country extends north to the broad valley channels of the Cimarron and the North Canadian rivers, although High Plains geology continues up the Central Plains into Nebraska and Wyoming. On the south the Southern Plains end raggedly and in spots sandily, where the Pecos River cuts into the far western escarpment of Texas' Edwards Plateau. On east and west the true Llanos country is equally well defined, on the one by the badlands and breaks bordering the Caprock Escarpment, on the other where the foothills of the Guadalupe and Sacramento and Sangre de Cristo mountains begin to heave out of the Great Plains. Add to this oval the Rolling Plains eastward to the 98th meridian and the country north to the Arkansas River, and one has defined the Southern Plains physiographic province as geographers do, some 240,000 square miles of it. But the heart of the Southern Plains, whether "heart" be defined viscerally or as true essence, lies deep in the recesses of the immense steppe plateaus that stand in the center of and at the northwest corner of the region. We still know them by their 400-year-old names: El Llano, or Las Vegas, for the sweep lying just eastward of the southern Rockies, and Llano Estacado (Staked Plain) for the great detached plateau spreading across the Southern Plains like jam on toast.

From atop the Llanos, where most of today's towns and few cities (Lubbock, Amarillo, Clovis, for example) are located and, hence, the perspective from which most modern residents and visitors form their impressions, the flat featurelessness seems to go on forever. One senses that somewhere beyond the ruler line of the horizon there must be an ending to it, but even in an automobile a crossing of the Llano Estacado from east to west takes nearly three hours. Eastward from the base of the Rockies a higher, gently rolling plain extends one to two hours by car before one reaches the lip of the Canadian Escarpment.

What is unique topographically about the Llanos is that its dramatic scenery lies below the general level of the country rather than standing upon it as in mountainous country. The Llanos is a land where high relief mostly yet awaits the unveiling process. But where exposed it falls away into the earth itself, a peculiar place of inverted mountains whose "peaks" are canyon floors, and whose sculptors are four rivers that sluice soil and rock apart to create the canyons and badlands country of the Llanos. This is a landscape formed not by tectonic uplift (although volcanic cones, like Capulin Mountain, and lava flows do exist on the Southern Plains) but by nearly 300 million years of erosion.

Except where a couple of state parks exist to protect them, or where the odd highway crosses them, these plains canyonlands are largely an unknown presence, even among most modern residents of the region. Only from the air is it readily apparent that along the eastern edges of the Llanos there is a series of sharply sliced gorges, tiered canyons, and blockily vertical badlands of rich coloring and an amazing variety of forms. The Texas Natural Heritage Program, an ecology project created by the state Parks and Wildlife Department and the Nature Conservancy, classifies this canyon country as a distinct ecological subregion, which it calls the Escarpment Breaks. In certain particulars these canyonlands seem the Southern Plains analogue to the Dakota Badlands, 800 miles to the north. But they are rockier, more vertical, more canyoned, in spirit more like a plains-scale version of the great canyons of farther west. The naturalist and literary critic Joseph Wood Krutch, searching for that mythical place where the landforms of the Southwest begin in *The Desert Year,* expressed what only a few have recognized. This is the place.

Red-tailed hawks and bald eagles leaving their summer homes in the Rockies to winter in the plains canyonlands sail over their first major Llanos canyon in New Mexico, where the Canadian and Mora rivers have routed a sinuous complex of canyons, 60 miles long and at places 1,200 feet deep, before emerging into eroded breaks below the Canadian Escarpment. Farther east, past Tucumcari Mesa, their buteo shadows undulate over the breaks of the Western Caprock, or Mescalero Escarpment, before smoothing out atop the Llano Estacado. With Tucumcari and Redonda mesas still in sight, migrating eagles pass over the trio of draws (Frio is the longest and most southerly; Tierra Blanca and Palo Duro are the other two) that unite 100 miles to the east to form Palo Duro Canyon. From these embryonic draws Palo Duro becomes a 45-mile-long canyon that reaches a depth of 800 feet and 12 miles in width before its opposing walls swing apart to form the Eastern, or Caprock, Escarpment.

Palo Duro is the main canyon of the Red River, but there are several others. Immediately to the north along the escarpment is Mulberry Canyon and beyond it the sand-filled valley of the Salt Fork of the Red. Wind-drifted sand softens and inundates the escarpment line beyond, so that of the Red River drainages that come off the Llano here only McClellan Creek furrows deeply enough to create a rock-walled canyon. But hawks or eagles heading southward along the Caprock fly over a series of little-known but incredibly beautiful canyons wending down through the strata of the Llano Estacado, canyons formed by the lower tributaries of the Red and by the Brazos and Colorado rivers of Texas.

Along the south wall of Palo Duro an eagle's-eye view would reveal three canyons so large and distinctive as to have identities separate from Palo Duro. North and South Cita canyons are two. Tule Canyon, a slit 700 feet deep, only a half mile across at the top, and my candidate as the undiscovered Yosemite of the Southern Plains, is the third. Immediately below Tule are the brightest and most castellated pair of canyons on the Llanos, the Caprock Canyons. Formed by the Little Red River,these are vermilion slickrock canyons, with vertical spires and turret walls, that seem to have been plucked out of the Four Corners country and set down in West Texas. Twenty miles farther south the serpentine, gray sandstone trench of Los Lingos Canyon coils through the junipers. Together with Quitaque Canyon a few miles below, it holds the headwaters of the Pease tributary of the Red.

Along the remaining 140 miles of the Caprock Escarpment, four additional canyons carry the waters of the Brazos and the Colorado off the Llano Estacado. Two of the longest draws on the Llano, originating in eastern New Mexico, are Running Water Draw and Blackwater Draw, sand-filled remnants, one senses when crossing them, of something that was once much grander. Fifteen or twenty miles from the Caprock, these link with the headward-eroding White River and the North Fork of the Double Mountain Fork to create a pair of twin canyons, grassy valleys hemmed in by rock walls and threaded by small, rapid streams. Blanco and Yellow House canyons are two of the three major canyons at the origin of the Brazos. The third is the Double Mountain Fork Canyon of Garza County. Along with the upper Colorado River's Muchaque Valley, twenty more miles down the escarpment, it is the

The Llanos, or Southern High Plains

most southerly and desertlike of the Llanos canyons, with geology and natural history different from those of any of the canyons to the north.

These fifteen canyons are my country. No one else seems to claim them, so I will.

ACCORDING to geologists the rocks can speak, and these are old rocks, with an amazingly long history written in sandstone cliffs and multicolored clays, pedestals and hoodoos and slides, fall lines and fossils. The story bespeaks of oceans and mud flats and erosional debris aprons sweeping eastward from the Laramide Uplift. Later, there are fill channels and pirating rivers and a panoply of species chasing and eating and mating with one another, although we may want to skip some of those details. But the geological details can't be skipped; the Southern Plains canyonlands are impossible to understand without them.

In field geology, a simple rule always works in country where the strata have been laid down as sediments: the stuff on the bottom is the oldest, that higher up is younger, and that on top is the youngest. Ascending a canyon like South Cita Canyon from its mouth to its head thus involves a paradox. While you know intellectually that as you ascend you are moving forward in time, the gut intuition is one of entering more deeply and intimately into the earth with each step.

Those dualisms of human nature, the yang of intellect and the yin of emotion, are not always at cross purposes when one endeavors to grasp the effect landscape has on the human psyche. Confronting a fossil embedded in ancient rock, even religious fundamentalists must feel some secret but quickly repressed tug of intuition about the continuum. Their creationist explanation for the amazing diversity of life on Earth stands as a monument to the human imagination; geologic and paleontologic evidence is testimony to the marvelous creative powers of nature. While science may have muted the role of the gods, it has not deadened our emotion or awe at what we discover. This acknowledgment of the *mysterium tremendum et fascinans* is about as close as we humans have come to grasping any underlying reality in nature, whatever permutations religion, philosophy—or science—might have put us through in search of something more.

WESTWARD over the horizon the Rocky Mountains are an unseen but sensed presence. Even with winter mirages, which bend light downward around the curve of the earth and make it possible to see, say, the lights of Hobbs from Lubbock, 100 miles away (the probable explanation for the famous "Lubbock Lights" of the early 1950s), it is not likely that one morning I'll climb the cliff behind my house and see the southern Rockies' Hermit's Peak plain as day, looming above Tucumcari Mesa. Visible or not, the Rockies are a major part of the story. But even they are not the starting point.

Geology loves a canyon the way all the world loves a lover. Canyons, like road cuts, show what a country is skeletally, its flesh sloughed off by wind, water, time. The canyonlands of the Llanos are the textbooks of Great Plains earth history, the "first book of genesis" as a National

Park Service scientist once put it, of the geology of the American West. Six major geologic periods—the Permian, Triassic, Jurassic, Cretaceous, Tertiary, and Quaternary—are exposed in the walls of these plains canyons, crafting in detail a record of more than 260 million years. To those who lack an interest in what these rocks and clays and shales can tell, the canyonlands are merely a beautiful but confused jumble of colors and sculpting. Understand even the elemental story and what had been just landscape becomes catnip for the imagination.

South Cita Canyon, one of the major tributary canyons of the Palo Duro system, is as good as any for getting a handle on what one is seeing on an ascent of a plains canyon. Start at its mouth, as Bill Brown, director of the Center of the American Indian in Oklahoma City, and I did in late October of 1987. South Cita can't tell us the tale entire, for two of the six geologic periods are missing from the strata of the Palo Duro system. But the two oldest periods and the two most recent ones are here. And the canyon-forming process on the plains is as easy to comprehend here as anywhere.

South Cita branches sharply from lower North Cita Canyon through walls of brick-colored clays that are the oldest soils on the Great Plains. The deposits are unmistakable: they make up the dazzling Red Beds of Oklahoma and North Texas and the characteristic bare terra-cotta soil of the Breaks just east of the plains canyonlands. Geologists call the system the Permian, and they call this formation of it the Quartermaster. Usually the Quartermaster is a friable clay, but it can form rock, actually shales and mudstone. Like most formations it is far from uniform. Look closely and there are gray-green circular spots, particularly where water washes over, known as reduction halos. Most noticeable of all are the white, usually horizontal veins of gypsum that interlace the Permian shales like latticework. The stuff is a form of salt, and anywhere you come across it in the canyonlands you'd better have drinking water along, for as soon as the pure spring waters from higher up reach the Permian the water becomes "gypy" and undrinkable.

So . . . the terra-cotta shale and clay with gypsum of the Quartermaster Formation are widely visible near the mouths of most of these canyons. Pretty. And mind-boggling. Stand amidst it and you're on the base rock of the Southern Plains; crumble it between your fingers and you're holding material that existed before the supercontinent of Pangaea began to break apart, before the Atlantic Ocean existed, before reptiles and cycad plants evolved. When the material exposed in the mouth of South Cita Canyon was deposited, the Southern Plains was a mud flat and most of this area was part of a detached sea, drying up in an arid climate and precipitating its salt into gypsum veins.

A half mile up canyon, South Cita Creek makes a sharp bend to the west against a dramatic Permian wall. The stream bed becomes rockier, with many cobbles and gravels. A hundred feet above the stream on both walls, pedestal rocks—six-foot columns capped by shale—provide the first hint of something stratigraphically different. South Cita is entering its Triassic section.

The Triassic is my favorite among the strata in the Llanos canyonlands. It was laid down by streams washing eastward from the ancestral Rockies, a mountain chain subsequently eroded flat, although along the present Front Range the ancient rocks were thrust upward by the rising of the modern Rockies and today form detached slab mountains, the Flatirons near Boulder, for example. The Amarillo Mountains, an ancient arm of the Appalachian Uplift long since buried beneath plains sediments, also contributed to the regional Triassic. The highest and lowest parts of the Triassic are greenish gray and lavender yellow shales, but most of the Triassic formations are sandstones, heavy, blocky walls that sheer in vertical sheets under erosion. The Triassic formations make the predominant high cliffs in the Llanos canyonlands, particularly in the canyons on the Texas side. The Triassic shales are responsible for the vivid, saturated bands of color that stripe the canyonlands. But without the sandstones, although this country would be an imposing badlands, it wouldn't be a true canyonlands.

In South Cita and the rest of the Red River canyons, there is what geologists call an "unconformity" between the Permian and the oldest Triassic, meaning that certain Triassic strata are missing. The plains Triassic levels that are present all belong to what is known as the Dockum Group, which includes both the Tecovas and the Trujillo formations, the latter occurring atop the former. Geologists recognize a number of different strata in the Dockum Triassic, but several impressive visual ones are especially worth knowing about.

As one climbs past the pedestal rock section of South Cita, mounds brightly variegated in lavender, yellow, and chocolate appear exposed on both slopes. In this part of the Palo Duro system these mounds occur almost halfway up the main walls. Upcanyon, in the state park, they are just above the level of the valley floor, where they comprise the part of the so-called Spanish Skirts, named after those that would cling to an Iberian maiden from waist to knee. These are the base shales of the Tecovas, the most eye-catching colors in the middle Palo Duro system, although Tule Canyon and lower Yellow House are about the only other canyons where the formation is prominent.

Above the banded Tecovas shales, another interesting strata crops out from the walls of canyons like South Cita and Tule, a jointed white sandstone that fractures into what looks like alabaster organ pipes like those of Bryce Canyon in Utah. Poke around in the Tecovas and you'll find any number of intriguing things: petrified wood from ancient swamps, the teeth of gigantic crocodiles known to paleontologists as phytosaurs, crystal-lined geodes, jaspers from which the native peoples crafted beautiful maroon spear and arrow points. In South Cita it is impossible not to rummage around in the Tecovas, since the creek slices directly through a massive Tecovas mound.

But you'll do some major boulder hopping first, because the creek has tumbled down gigantic blocks of sandstone from above, where it has broken through the Trujillo.

Erosion and gravity attack weakness. When they were laid down as sandbars and dunes, these sands were easily eroded. Now they stand as the most resistant major objects in the Llanos canyons, cliff-forming walls that (as in the Tule Canyon Narrows) stand 300 feet thick. In some places there are two sandstone ledges, in others as many as five; invariably, they are separated by beds of softer shales, which makes the Trujillo the outstanding fall-line formation of the Llanos canyonlands.

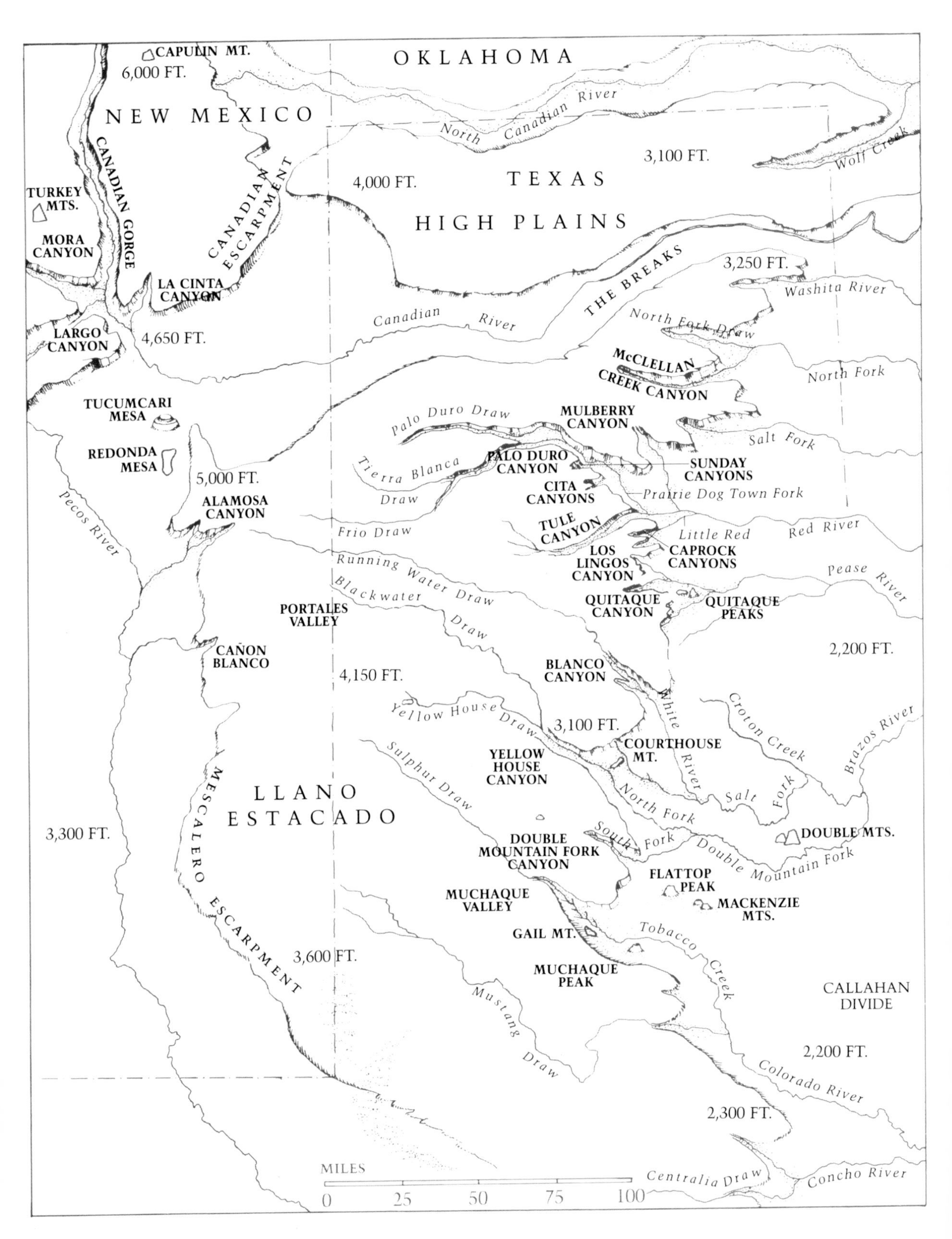

Canyonlands of the Southern Plains

Everywhere the creeks and springs cross these sandstones they have sliced away the underlying shales and clays much more sharply than the stone above. The effect creates overhangs and caves, and where there is water, pools and waterfalls of exquisite beauty, although none of the falls has a vertical drop of more than about 30 feet. This same effect, what geologists call differential erosion, makes the Dockum sandstones the premier pinnacle- and butte-forming rock in the canyons.

Their colors are gray, brown, greenish, and in certain spots—Little Sunday Canyon, the Tule Canyon Basin, Caprock Canyons, La Cinta Canyon at the mouth of the Canadian Gorge, the north wall of the Double Mountain Fork Canyon—a spirit-tugging vermilion, left as a legacy of a time when so much oxidation was taking place that the entire planet rusted to a Martian red. Lay a hand on it. It is cool to the touch, grainily smooth, sculpted into sensuousness by wind and water. It is red slickrock, the quintessence of the southwestern canyonlands from West Texas to Arizona and Utah.

The ascent up South Cita Canyon, through car-sized sandstone boulders, pools, and waterfalls, is far from over by the time you top out of the Triassic section, but the canyon rims are now in sight, 200 feet overhead. The geologic story seems almost over. Yet the Triassic Period ended 195 million years ago. The lower of the white rimrocks overhead is less than eight million years old. Obviously, something big is missing.

The explanation is another unconformity, an immense one. The ensuing millions of years, two whole geologic periods, are missing from the Palo Duro system. Perhaps they were never laid down here; more likely they were eroded completely away. But to grasp the Jurassic and Cretaceous periods, critical times when true birds and mammals evolved, the dinosaurs disappeared, and present continental separation occurred, it is necessary to turn to canyons where these chapters are not absent.

The Canadian Gorge in New Mexico is the Llanos canyon that preserves the best expression of the Jurassic Period. Near the valley floor where the Canadian begins routing its channel into the plains, and for 25 miles downcanyon, the Canadian has peeled away the overburden to expose smooth, reddish-white sandstone walls covered with desert varnish. The walls are Jurassic sandstone, laid down as sedimentary outwash from the highlands to the west. The formation crops out as well from base walls along the 1,000-foot-high Canadian Escarpment. And it forms a principal tier in Cerro Tucumcari, where it was first identified in the American West by a European paleontologist familiar with Switzerland's Jura Mountains, where the formation was discovered.

The other major deposition missing from the Red River canyons, the Cretaceous, is also found in the Canadian Escarpment and Gorge country. It contributes the Dakota sandstone that makes up the house-sized, blocky rimrocks. As the great Cretaceous Sea, extending from the Arctic to the Gulf, gradually withdrew at the close of the Mesozoic Era, around 70 million years ago, it must have left sedimentary deposits, full of bivalves and gastropods and reptile bones, all over the West. But on the Texas side of the Llanos the only canyons that expose Cretaceous rocks today are those on the bottom end of the Caprock Escarpment. Yellow House Canyon has cut deeply enough into the plain barely to dip into a white Cretaceous ridge. But the best and most intriguing Cretaceous rocks in the Texas Llanos canyons start on the south rim of the Double Mountain Fork Canyon. From there southward the rimrocks of the Caprock Escarpment, the Muchaque Valley, and the upper rims of mesas like Flattop Peak (the Texas-side twin of Tucumcari), Gail Mountain, and Muchaque Peak are chalky Cretaceous limestone.

The Cretaceous Period and the entire Mesozoic Era ended abruptly 65 million years ago in a great crash of extinctions, the possible result of an asteroid impact whose dust cloud obscured sunlight long enough to kill off most plant life. The ensuing Tertiary Period has left only the fragment of a record in these canyons. But that fragment, spanning just the last five million years of a period that lasted more than 60 million, deposited the signature rocks of the Great Plains from Texas to South Dakota and Wyoming. In South Cita—in fact, in every one of the Texas canyons northward from the Double Mountain Fork Canyon and around the entire escarpment perimeter of the Llano Estacado—the white or pinkish rimrock tiers at the tops of the canyons and escarpment cliffs are that record. They look Cretaceous, and some of the early geologists who worked the Southern Plains thought they were. But these cliffs are Pliocene "caliche," a kind of arid-land limestone formed when subsurface moisture containing calcium bicarbonate evaporated upward, creating a rocky, mineralized crust on the surface, the famous "caprock" of the plains.

Bill and I didn't ascend South Cita Canyon all the way to the caprock that day in October. We stopped within sight of the rimrocks but short of the church camp in the grassy reaches of the upper canyon. No need to explore farther. Any halfway adventurous soul who has been into the shallowest canyon or draw on the plains knows what rotten, friable caliche caprock feels like, for every draw that cuts deeply enough to form a canyon slices through the crust rock. And although it is always startling to mount that last rim and find yourself standing on the surface of the plains, we had to turn back. The Jeep was miles away.

Common as they are on the plains, the Tertiary and Quaternary surface formations that comprise the canyon rims are hardly boring. The roughly textured caliche walls are the most eye-catching, but the so-called Ogallala Formation includes red clays and silts, gravels, sandstones, even lenses of volcanic ash from the fireworks that were lighting the night skies of the Pliocene. This is the Rocky Mountains you're standing on, never mind that the peaks where this debris originated cannot be seen from here or that the clash of continental plates that set the whole shebang in motion took place 2,000 miles away. The surface of the plains is an erosional apron, fanned eastward as water and wind wore the rising, modern Rockies away and great rivers carried the effluvium toward the Mississippi.

Down where I live, in Yellow House Canyon, big earthmovers chew away at the canyon to get at gravels that began their journey to West Texas above timberline in the San Juan Mountains of Colorado. When those gravels were being washed eastward, the Brazos River was probably the greatest river on the Southern Plains, the southern version of the Missouri. We're still dependent on the snow melt that surged through its channel and those of the other ancient plains rivers, for huge quantities of it saturated the sands and gravels and still sit in the sandstone tub of Triassic rocks. We call it the Ogallala Aquifer.

This Ogallala Formation is also fossiliferous in the more traditional sense. In the 1890s this is where many of the great paleontological expeditions emanating from the Smithsonian and Harvard and the New York Museum of Natural History were bound. The Blancan Age takes its name from Blanco Canyon, where fossils that are diagnostic for this time period were first discovered in its framing white cliffs. Yellow House and Quitaque canyons have significant paleontological sites of Illinoisan Age fauna. Sunday Canyon and Tule Canyon, and especially North Cita Canyon, all have famous Pliocene sites.

What the Rockies outwash captured in stone is a piece of a long-gone time when grasses had overspread the plains and circular lakes called playas captured rainfall in a rudimentary form of drainage. The climate was growing more arid but was wetter than today, and crocodiles and giant tortoises existed alongside ancestor species of grazing and browsing zebras, camels, and antelope. Saber-toothed cats and bone-crushing dogs preyed on these beasts, and ancestors of the modern coyote already played their jackal role as scavengers. The linking of North and South America provided the possibility for faunal exchange from the southern latitudes, hence ground sloths and giant armadillos. Such was the Southern Plains one to three million years ago, about the time the Red River began to cut the Palo Duro Canyon system.

Not a bad story, for mere rocks.

THE BAND of hunters had been camped along the river in a wide, rock-walled canyon that sliced across an immense, evergreen-dotted plain for nearly a moon cycle when Eru'hu finished his hunting rituals. For three sunrises Eru'hu had sat in a chilling drizzle on a flat-topped mountain in the canyon, fasting and chanting his songs to Kuna'pele-le, the god and keeper of the great hairy elephants. He had pledged strictest observance of all taboos and abiding respect for these red giant people if one would allow itself to be killed so that Eru'hu's people might eat and thrive. For three days there had been no sign, but this morning a raven had flown low overhead, dipping its wings. It had croaked once and then veered off directly upstream. Eru'hu had smoked to the four gods in thanks and stood stiffly, giving the sign to the camp of dome-shaped lodges below, and the hunters had begun to prepare their spears. By the time Eru'hu had climbed down from the cliffs, the women had stripped the hides from the bone and pole frameworks and the band was preparing to move, its scouts far in advance, the hunters singing songs of confrontation and preparing themselves as if for battle.

The shaman's medicine was good. Two sunrises later, the sky now clear and the grass drying, the scouts returned in a state of great excitement to report the presence of a small herd. Five adults and two juveniles of the red giant people had been seen ripping up the tall grass in great bites near a marshy lake where two shallow draws came together to form the head of the canyon. A-wagh'hek, the band's most accomplished hunter and warrior, tested the wind direction and sent three fire men in a wide circle beyond the elephants with instructions to fire the grass as soon as the dew had gone. Then he arrayed the remaining hunters, eighteen in all, in the lush stream-bank vegetation below the marsh.

They caught the acrid smell of burning grass even before they saw the smoke and heard the shouts of the young ones who had gone out as beaters. As A-wagh'hek had known she would, the shaggy cow with the broken tusk who led the red giant people sought to escape the flames by taking the herd up to their hairy bellies in the marsh. A-wagh'hek gave the signal and the hunters swarmed from their hiding, their shrieks lost in the trumpeting of the elephants. The Great God Kuna'pele-le often demanded sacrifices when she surrendered up one of her elephants, and she did this day. The first hunter to enter the marsh, a careless young man barely beyond his puberty rites, became mired in the soft mud and at once disappeared under the trampling of the enraged bull. But the mud slowed and tired the red giants until the hunters could finally isolate the two juveniles, which fell to a trick: a daring hunter would approach from the rear and grasp a hairy tail, and when the animal turned in a spray of mud and water, two others plunged their bright spears deep into the exposed soft spot behind the foreleg. Wheezing away their lives, their trunks spattered with frothy lung blood, first one juvenile and then the other toppled slowly into the lake. When the sun was midway in the morning sky the adult elephants finally gave up their efforts to rouse them, although one of the younger cows returned to the bluff over the lake again and again.

After sunset and a day of butchering with flint knives and hand axes, Eru'hu watched the red line of the fire, now burning far to the north, and, gorged on charred fresh meat, he dreamed a strange, disturbing dream of failing medicine and of searching, searching without being able to find the red giant people anywhere.

WE CALL THEM the Clovis hunters today because the first discovery of their finely knapped fluted spearpoints embedded in mammoth remains was made near the town of Clovis, New Mexico, in what was once a marsh on Blackwater Draw. No one has the slightest idea what they called themselves, what names they gave the Llanos or its animals. About all we are likely to know is that they were here, that they were probably (although not certainly) the first men and women on the Llanos, and that they arrived from Asia with a remarkable flint tool kit. Like their earliest ancestors they were preeminently grassland creatures. Some archaeologists call them and their successors "Llano Man" because the greatest number of their ancient camps and kills have been found in the draws and canyons of the Great Plains. In fact they seem to have spread over much of North and South America—in not much more than 300 years, if biologist Paul Martin is right—about twelve centuries ago.

They found a plains country that had changed from that of the Pliocene, one more like the Llanos we know today, yet at the same time wondrously different.

Something had happened to the apron of debris that had swept smoothly down from the foothills of the mountains. It had become dissected. In late Tertiary and early Quaternary times, in response to another collision of the ever-shifting continental plates, the Sangre de Cristo Mountains had lifted amidst a spectacular pyrotechnic display in the West (more than fifty volcanoes were active at once). For a time the Brazos had headed in the Sangres, probably in the drainage now occupied by the headwaters of the Pecos River. The upper Canadian was born out of the snowmelt of the Sangres and had begun to eat out its canyon, but for centuries it was merely a tributary of the Brazos.

Things were happening on the eastern edge of the apron, too. By the middle Pleistocene, a million years ago, headward-eroding streams fed by rainfall and by the springs leaking from the Ogallala Aquifer began to cut westward into the Llanos, sometimes following the valleys of old rivers, often simply eating their way from one playa lake to another. They were our modern Llanos rivers, and during wet pulsations they began the process of carving their canyons, a land sculpting that is still at work and as recent as the last rain shower. As the canyons were cut downward, more and more strata were exposed to weathering and differential erosion, and gradually the cliffs began to recede until eventually the walls of one canyon intersected those of the next. By the time the Clovis hunters arrived, hydrology had not only produced westward-migrating canyons, it had effected a 200-mile recession of the eastern escarpment. Since the arrival of humans the scarp has retreated another mile.

Then, about 90,000 years before the Clovis folks, the two oldest of the headward-eroding rivers had torn through the soft rocks of what is now eastern New Mexico and in the space of a few thousands of years isolated the Llano Estacado, making it a plateau. At roughly the same geologic instant the lower Canadian ate far enough to capture the upper Canadian just below its gorge, and the northward-eroding Pecos managed to snag the waters of the main stem of the upper Brazos. The once mighty Brazos now headed on the plains, its ancient Rocky Mountain channel still visible in the Portales Valley and the fishhook-shaped upper valley of Alamosa Canyon. Now mountain rivers, the Canadian and Pecos rapidly effected dissection, creating the Mescalero and Canadian escarpments and buttes, badlands, and mesas in between.

Physiographically, it was the modern Llanos the early hunters saw. Not so environmentally. Perturbations in the earth's orbit produced major climatic swings during the late Pleistocene. Seventeen major cold-warm pulses, although they never sent ice sheets onto the Southern Plains, enormously affected Llano lifeforms. Not only did the glacials link North America with Eurasia via the Bering land bridge, but also the cool, damp conditions south of the ice made an ideal habitat for Eurasian grazers. With a summer temperature that probably averaged about 61°F, eighteen degrees cooler than today, even Rocky Mountain vegetation—blue spruce, firs, ponderosa pines, Rocky Mountain cedars—established colonies on the Llanos.

Ponderosa pine in the Dakota Cretaceous rimrock of the Canadian Gorge

The pulsing climates had created thousands of habitats, and mammal evolution had radiated like never before. The Llanos swarmed with exotic animals. "We live in a zoologically impoverished world," Alfred Wallace wrote in 1876, "from which all the hugest, and fiercest, and strangest forms have recently disappeared." He was talking about the Pleistocene, the mammal skeletal remains of which first became known in the late eighteenth century.

They are gone now, some thirty to thirty-five genera of animals that were right here on the plains as recently as 10,000 years ago: two kinds of giant sloths; a beaver the size of a modern black bear; a peccary as big as a European boar; at least eleven species of antelope, several with four horns; scores of species of horses, the taxonomy of which is still confused; a camel a quarter-again larger than the modern camel; short-faced bears; the dire wolf; Smilodon, the sabertooth that killed young mammoths by stabbing them; Panthera, the giant steppe lion; mammoths and mastodons and giant bison; big scavenging teratorns and condors and eagles.

They are gone. In many instances the plants they fed on are still here, like the caliche globe mallows found in the fossil excrement, or coprolites, of the extinct sloths, the cottonwoods and hackberries once shredded by edge-loving mastodons, the Stipa grasses that coevolved with horses. The African Veldt yet has some eighteen grazing species; its Pleistocene animals largely survived. The American plains were left impoverished, with only six large grazers and browsers, all of them except deer and pronghorn Eurasian immigrants. This explains why there were so many bison, a dwarfed, weed species whose population exploded with all those niches vacant. And, as Alfred Crosby has pointed out in *Ecological Imperialism,* the lack of native animals was a key element in the vulnerability of the New World to Europeans, with their complex of domesticated plants and animals that thrived in disturbed ecosystems.

The Pleistocene extinctions were a major ecological disaster, unquestionably the most serious one on the plains in human times. Moreover, and this takes me into the eye of a controversy that has flared for a century now and presently is an academic fire storm, it was likely the first of a series of human-caused ecological collapses on the American plains.

Extinctions are a natural consequence of an evolving earth. There had been great extinction episodes before, and given the radiation and specialization of lifeforms during the Pleistocene, extinctions were inevitable. Some of the Pleistocene beasts—the dire wolf, all those antelope, for example—clearly lost the competition to their modern descendants. But something is naggingly wrong about many of those extinctions. If the warming climate was to blame, why did only large species become extinct and not small ones, as had happened in earlier extinction episodes? Why didn't the South American immigrants to the plains thrive? Why were the bulk of the survivors Eurasian rather than American species? Why weren't there evolutionary replacements for the vacated niches? And why did all the extinctions occur almost simultaneously (biologically speaking) about 11,000 years ago?

Paul Martin of the University of Arizona and a dozen or so very bright colleagues and students think they know the answers. Spend a few afternoons with *Quaternary Extinctions*, the hefty tome Martin co-edited with Richard Klein in 1984. Sift the ideas around, pro and con. It is hard to escape the conclusion that the early hunters, particularly the Clovis people but very likely the Folsom and Portales people who follow them in the archaeological record of the next 2,000 to 3,000 years (and whose specific cultures are identified by spearpoints with names like Plainview, Plano, etc.), were seriously, perhaps critically, involved in the domino effect of the extinctions. That conclusion is deeply disturbing for those of us who have looked to tribal peoples and animistic religions for solutions to living in balance with nature.

There is absolutely no question that between 11,000 and 8,000 years ago these Paleo bands hunted most of the grazers that subsequently became extinct. On the Llanos their points have been found in the remains of mammoths in Blackwater Draw and at the head of the Washita River, near Miami, Texas; as well as in the remains of extinct bison in Blackwater Draw, in Yellow House Draw at the Lubbock Lake archaeological site, in a canyon near Folsom, along the Mescalero Escarpment near San Jon, at the base of what was once a cliff in Running Water Draw, in Tule Canyon, in one of the Caprock canyons on the Little Red. A main base of the Folsom hunters, who specialized in the giant *Bison latifrons*, seems to have been near present Abilene, but one of their religious shrines, the skull of a giant long-horned bison set facing the sunrise on a sort of bone altar, was in one of the redrock canyons of the Little Red.

They worshipped the animals they hunted; they also watched them disappear one after the other, not only the ungulates but finally the predators that had depended on them. The extinction scenario, on a computer model of radiocarbon dates, portrays the vortex of the phenomenon being here on the Llanos, then diffusing outward like ripples on a pond until every last animal was gone.

It may have happened like this: the warming climate had reduced to isolated relicts the suitable habitat for many species; the American-evolved animals had emerged in an environment lacking human hunters, thus, had few defenses; with very long gestation periods and a probable hunter preference for females and juveniles, their populations were unable to rebound; the extinctions happened quickly and in local isolation so that the Paleolithic shamans either misunderstood or could not grasp the implications: that their way of life, their way of using the plains, was over.

WHETHER the succeeding occupants of the plains learned from this lesson—indeed, whether they even knew of it—is not apparent. On the face of it, those who followed the hunters did not wipe out the surviving plains animals, although they may have pressured bison toward smaller size and shorter gestation times. As a matter of archaeological record, except in a few well-watered canyon glens like the Lubbock Lake site, both bison and humans may have all but abandoned the Llanos during the great drought known as the Altithermal, one of the inverse climatic swings that seems to have hit the Southern Plains around 7,000 years ago. For nearly 3,000 years the Llanos were parched

CLOCKWISE FROM TOP LEFT: *A hawk's-eye view of South Prong Canyon, Caprock Canyons State Park; Blanco Canyon; Capitol Peak and the Sunday canyons, Palo Duro Canyon State Park; Spires and fins topped by Dockum sandstone of the Triassic Period, Caprock Canyons*

and sand whipped. The bison range shifted to the east and west of the plains. Then, about 4,000 years ago, the cycle swung back. Following the canyons westward, both large animals and Archaic gatherer-hunters gradually reoccupied the Llanos country.

Hundreds of these so-called Archaic sites exist on the Southern Plains, but so far no one has conceptualized them into a chronology, that pasttime so dear to the heart of the archaeologist. But evidently the Archaics were enamored with the canyons. If the pattern associated with Little Sunday and Tule canyons holds, they liked to place their base camps on the rims overlooking the canyon floors, perhaps so that foraging parties could be watched. In Tule the number of their camps increased greatly downstream from the Narrows. Perhaps the deep gorge itself was a sacred spot, one not to be profaned by ordinary activity.

They were all over these canyons, perfecting a canyon lifestyle based on exploiting the full range of the local environment and on a low population. Not that they moved so very lightly over the landscape. With atlatl-propelled darts and spears they killed bison they had stampeded off cliffs and herded into arroyos, and several sites indicate that only select animals and select portions were taken. Unquestionably they burned, not so much broadcast fires to promote grass growth and aid hunting, but as plant gatherers do, to create a mosaic of species. Yet we have no record of a massive ecological disequilibrium caused by the Archaic Meso-Indians, and it is hard to resist the conclusion that the reason was that these people did not specialize as the Paleo hunters had. They adapted by diversifying.

Theirs was probably the most successful adaptation human ecology has so far made to the Southern Plains. Picture their lives in the mind: lying on a sunlit slope watching a herd of animals below, every muscle, every sense operating at the level evolution prepared us for; grinding seeds by a singing canyon waterfall surrounded by friends and relatives; sitting on a rimrock at night, watching dim, dancing forms in the fires on the far canyon rim and hoping for meaning from the glittering panoply overhead.

In the last 1,500 years before Cabeza de Vaca and Coronado, the native peoples had begun to experiment with other ways to adapt to the Llanos. From both east and west, groups with the Mexican cultivar complex of corn, beans, and squash began to enter the Southern Plains. Probably those from the east were colonizers from the great Mississippian cultures, known on the plains as Plains Woodland and Plains Villager peoples; those from the west were probably settlers from the irrigation cultures that had taken root in the Far Southwest. The easterners, possibly Caddoan speakers, moved up the rivers; the westerners, maybe Jornada groups of Mogollon background, looped into the canyons from southern New Mexico.

At least some of them built a small, sandstone slab pueblo in the Tule Canyon Basin, but their major plains pueblos, represented by the Antelope Creek ruins and several others, were in the Canadian Breaks of the Texas Panhandle, yet not in the Canadian Gorge (an argument for their being easterners rather than New Mexicans). The horticulturalist mode of adapting was to push deeper into the interior of the Great Plains during times of good climates and to fall back toward the moist woodlands when the droughts hit, as they did around A.D. 1299 and again in A.D. 1450. The sensible strategy of abandonment, in other words, one recognizable to a Dust-Bowler of the 1930s or the Llanos farmer of the 1980s. The Jornada folk did something equally interesting. They stayed, but seem to have fallen back on the old Archaic lifeway when times got bad.

Some of both easterners and westerners must have still been around when, in the late fourteenth century, an aggressive new intruder arrived to disperse them. The intruders came down from the Canadian northland in large numbers and with many wolflike dogs as beasts of burden. They had a conviction of their own superiority and a desire to take over the bison and the trade and the country. One of their divisions swung west of the mountains to become the Navajo. The eastern division found canyons like Palo Duro and Blanco much to their liking.

They would come to be known in plains history as Querechos, Teyas, Lipans, Llaneros. They were the Apaches.

CLIMBING BACK out of Palo Duro Canyon, Bill and I ascended Indian Point Trail, the very trail the U.S. military used in 1874 to ride down on the last big Native American village ever to be set up in the Llanos country. It was half an hour before sunset; the canyon behind us was almost a melodrama of autumn shading thrown into high relief by horizontal sunset light. Cotton-ball clouds, extending away until they merged with the encircling horizon and all laid out on the same elevation, the way clouds often hang over the plains, dotted elongated shadows across the land.

All that geology: millions of years of Earth time opened like a book by flowing water. All that human continuum. How many generations—two hundred? three hundred?—wooing and loving and, yes, sometimes hating this country, but sunk into it up to our armpits.

We sat on the rim and looked at it until the stars came out.

Chapter 2

Song of the Desert

Most of the Southwest's frontier explorers, hunters, ranchers, and cowboys were southerners, with east Texas heavily represented. . . . Such men were almost fanatical in their mission to rid the southwestern rangeland of wolves, bears, and other varmints.

—David Brown, *The Grizzly in the Southwest*

Throw into the fact barrel, finally, a 1983 report from the Council on Environmental Quality concluding that desertification . . . proceeds faster in the western United States than in Africa.

—Wallace Stegner, *The American West as Living Space*

You gotta leave it better than you found it.

—Rancher Riley Miller

IT IS A HOT, glaring noon in late August 1986. Fledgling Mississippi kites are cutting arcs in the blue-white sky, awkwardly pursuing the few straggling cicadas. Windows down in my truck to suck in a sweat-drying breeze, I am hewing a dead-straight line across the ruler-flat Llano Estacado, en route to the entrenched upper canyon formed by the Double Mountain Fork of the Brazos River. Buffalo hunters, I have heard it said, determined their riding distance across this plain by counting the hoofbeats of their horses from one camp to another. Navigating the awesome expanse of buffalo and grama grasses—like nothing else in nature quite so obviously as a becalmed ocean, the universal metaphor since ocean-going people first saw the Llanos—early travelers arched arrows in an overlapping pattern or drove stakes in the ground to keep their way. Since the Clovis hunters, humans have also learned to allow for "drift" when crossing the Llanos, for the imperceptible slope from northwest to southeast would drag you off a direct line. Those who failed to correct for it and missed springs or the few permanent playa lakes became casualties of Llanos natural selection.

With their compasses and theodolites, their geometry and their trigonometry, the engineers who laid out the highways and farm-to-market roads on the twentieth-century Llano Estacado had no such problems. These are among the straightest highways on the continent, and in the heat mirages of summer they carve tangents that seem to rise and vanish into the shimmering heavens.

Mirages, like rainbows, aren't supposed to be there when you are. But the true phenomenon is not so much invention as distortion; although objects can be elevated, bizarrely magnified, even turned upside down by summer mirages on the plains, the objects nevertheless exist. Mirages on the Llanos are most associated with the visual distortions created by draws and canyons, and old cowboy Rollie Burns claimed that on a cold and clean December morning in the 1880s a "perfect mirage"

OPPOSITE PAGE:
Rocks that communicate across the centuries, Comanche petroglyphs from Cowhead Mesa

enabled him to see into the bottom of Yellow House Canyon while encamped eight miles away from it. Of course that was a winter mirage, which bends light down rather than upward. So today there are no canyon's blue mesas shimmering high above the baking plain. Instead, I glide the truck to the very rim of the Double Mountain Fork Canyon where the ground yawns open startlingly.

Judging from the descriptions in the diaries of the explorers and early travelers on the Llanos, reaching the rim of one of these canyons has always been a moving experience. It still is, maybe more so now that the surface of the plain is so transformed. Where a century ago the plain itself, the "dread Yarner" as the buffalo hunters called it, represented the wild and unknown, these days the canyons play that role, and the Llano is so tamed and civilized as to hardly seem a part of the West at all except for its skies.

It's a base reversal of extremes, almost like switching centuries. One minute you are on the plain, still vast but its meditative qualities now sacrificed to criss-crossing power poles receding into the distance, the grama grass and catclaw acacia having given way to row after row of cotton or sorghum, at night mercury vapor lamps by the hundreds polluting once-brilliant West Texas nights. The next you are crunching short yellow buffalo grass underfoot at the brink of a great rent in the plain, the air suddenly perfumed with sunned juniper and sagebrush. Below, between mesas and badlands, are snatches of the sandy coilings of a river. Overhead, hawks spiral in the thermals. At your back, the twentieth-century experiment; below, the ancient landscape, chipped away in places, but its ecology, its sacred circle, still intact enough for meaningful journeys.

This afternoon I follow the ranch road along the north rim of the upper canyon until I can ease the truck down into the gorge by way of a road used by pumpers and hot oil trucks to reach the Garza County oil field a few miles downriver. My destination is the opposite direction, up the narrowing canyon to Cowhead Mesa. I plan to hike up the river into Mooar's Draw, spend a night, and then loop back to the truck the next morning, a short, leisurely walk through the canyon provided I cover some ground this afternoon. That may not be easy. I'll leave the truck in the shadow of Cowhead Mesa, a landmark humans have never easily ignored. Most of the mesas and buttes in this country beg exploration, but this one is more compelling than most. On the sheer Chinle sandstone wall that makes up its west face, Cowhead Mesa preserves the best panel of rock art found anywhere along the Caprock Escarpment. Its incised figures, or petroglyphs, have arrested my exploration of this canyon more than once.

I park the truck on a sandy bluff over the river, open the camper shell and watch my male Siberian husky, Tule, bound away to make a rapid reconnaissance. I lock the truck and shoulder my pack, and within half a dozen strides the mesquite, junipers, and head-high cholla cactus have closed in behind. I have with me the standard sort of backpacking gear: a light down bag and pressed-foam mattress, a rain poncho but no tent. I have a camera, a couple of lenses, and a light tripod—my bulkiest items—a topographical map, but being an advocate of the so-called Australian method of handling snakebites, which basically means leaving the bite itself alone, no snakebite kit. I have granola bars and peaches and coffee, but mostly I have brought water, three quart canteens. There are springs in this canyon, and the water is clear, cold, and good. But they are back up in the side canyons, with the rattlesnake dens, and in August the low daytime flow in the Double Mountain Fork ups the chances for Giardia microbes.

So you need the water to hike this country in the summer. Already it is almost 90°F; by 5:00 P.M. it will be over 100° here in the canyon. Full exposure on such a day on the southern High Plains will dehydrate a 150-pound mammal at the rate of about a pint of water an hour. And of all the major canyons of the Llanos this one is the lowest in altitude (about 2,950 feet at its head), most southerly (33.06 degrees north latitude), and comes closest, in terms of both climate and ecology, to being a true desert.

I stop, wrap my hair back from my eyes with a bandanna, and look back through the brush at the truck, an alien piece of tin and plastic and rubber surrounded by the wild and natural. But it's a momentary illusion. Ranchers and oil-field trucks ply the two-tracks through this canyon, and on nice weekends dirt bikers race a course laid out with pink survey ribbons.

Across the sandy river channel Cowhead Mesa stands boldly, looking unruffled by it all.

COWHEAD MESA is a typical bread-loaf-shaped western mesa, indistinguishable at a glance from dozens of others in these canyonlands. Unlike the rest, it is currently under investigation by the National Register of Historic Places because of its Indian rock art. Compared to farther west, there is not a great deal of rock art on the Southern Plains, and except for the Cimarron Valley and Rocky Dell in the Canadian Breaks, almost all of it is in these canyonlands. A dozen or more minor rock art sites are in Palo Duro, Tule, Blanco, and Yellow House canyons, and like Rocky Dell, most of them seem traceable to those traveling artists, traders from Pecos and Santo Domingo and other New Mexican pueblos. The Rocky Dell pictographs show a horned serpent like those that appear so frequently in Galisteo Basin near Santa Fe, and on Yellow House Crossing Mesa a Kokopelli is etched into the sandstone. Kokopelli, the humpbacked flute player, was the Pueblo equivalent of the traveling salesman, a slippery rogue with a pack of goods who could sucker men and coax women into the bushes. This is his most easterly appearance on the continent.

Cowhead Mesa and the Double Mountain Fork country are intriguingly different for two reasons. For one thing, the canyon is the rock-art capital of the plains canyonlands, with nearly three dozen sites now known. A remarkable number of them also clearly tell historic-era Plains Indian stories, of priests and wagon trains and war honors. Cowhead Mesa is such a one. It seems to reveal an Indian perspective on a famous event in Texas history, the San Sabá massacre of 1758.

In the southwestern world of the 1750s, the pious Spaniards in San Antonio had been approached by the Apaches, the peoples their first *entradas* had found on the Llanos and the selfsame who had long raided their settlements and made fun of their priests, with a pleading request for a mission. The Franciscans interpreted this to mean that the Lord had at last delivered the pagans unto the fold. They built the mission about 100 miles northwest of San Antonio along the San Saba River. It was, in effect, a blood sacrifice. What the Apaches really intended was to intrude the Spaniards into the eye of a tornado sweeping down the Llanos upon them, a horse-mounted tornado of wild hunters from the north who would enter history as the Comanches. As the Apaches had probably calculated, almost one year after the mission was built a large and splendid force of Comanches shoved their way through the gates, killed two priests, and burned the buildings to the ground. Later they added humiliation to the lesson, routing a large Spanish army sent to punish them, capturing its artillery, and chasing the proud *conquistadores* who had humbled the Aztecs around the prairie like chickens.

If the rock art on Cowhead Mesa depicts what it seems to depict, one group of Comanches involved in the sacking of San Sabá headed up the Brazos River, into the remote canyon at the head of the Double Mountain Fork. There they carved into the sandstone face of the mesa their impressions of the event: mission buildings like layered cakes with crosses on top, flames licking through them, roundabout men wearing the frocks of priests and others with three-cornered hats, scenes of personal combat. One glyph is of a bovine with a long, ropey tail and spi-

Saturnian forms, badlands of the Double Mountain Fork Canyon

raling horns, perhaps an initial Indian impression of a Spanish longhorn. It's the figure that gives the mesa its name.

There are other figures, too: a bear paw, a turtle, illuminated tipis showing the frame poles beneath the covers, a large "shield" or "calendar" figure, a particularly nicely done mounted horseman with lance and shield, executed with a Picasso-like grace that still expresses the artist's intent to show that the horse was a fine Arabian. Most powerful of all is a looming figure in Comanche buffalo headdress, a shaman with outstretched arms much reminiscent of the ghostly Anasazi figures that hover over box canyons throughout the Four Corners country. And there are lines and squiggles whose meaning no one knows, although some scholars suspect that the world-wide distribution of such symbols points to phosphenes, patterns created by the human brain and "seen" behind tightly closed eyelids or released through dream, vision, or drug states.

A question that has long puzzled me is why rock art was done where it was. Why this particular mesa, for example, when dozens of other sandstone faces are scattered about this canyon? Practical archaeologists point to things like nearby water, protection from weather, and so forth. But something else must have been going on, what the geographer Yi Fu Tuan calls "topophilia," which was bound up in the native perception of sacred place and *genii loci,* the spirits that inhabit particular power spots in the landscape.

Nonsense maybe, but one culture's nonsense is sometimes another's religion. And as a matter of fact, I have been face-to-face with the spirit that inhabits a particular rock-art site on a tributary of the upper Canadian, over in New Mexico.

It was a still fall day, a little autumn haze in the air. I was with Katie Dowdy, who was doing the fieldwork for a master's thesis on plains rock art. Thoughtfully parking my old Fiat in a pasture where the horses could have a fair try at licking off the remaining paint, we had hiked upstream a mile or two to the site, a small hemispheric rock amphitheater on the south bank of the creek. Walking in we had passed excellent locations for sites, had seen nothing. But the pumice walls of this little grotto were covered top to bottom with dozens of stunning spirals, shields, nebulae, animals' spirits, and dream images.

Katie was out of sight below me. I was standing on a shelf by the uppermost petroglyphs, still mildly puzzling over why this particular spot, when the spirit appeared for the first time. I was framing a photograph when I heard the noise of it. Thinking a rattler was rustling in the grass behind me, I froze. The noise—easily audible, a steady whirring—grew louder. I turned around.

It wasn't a snake. Whatever it was, it was in midair, about 15 feet away from and slightly below where I stood at the rimrock. I registered an impression of Katie in the grass below, openmouthed. Our gazes converged on an indistinct visual disturbance in midair, a whirrrrring plains dervish that was marching in stately fashion around the circumference of the rock cove. Too stunned to speak or even gesture, we wordlessly watched it walk past us and down to the creek . . . where, once it hit the cottonwoods, it assumed the form of an ordinary whirlwind.

Rational, western minds can be satisfied, then. Twice more during the next hour the spirit appeared, followed the same path: emerged from the boulder corral a small whirlwind that, I would have argued earlier that day, was merely the creation of the peculiar topographic qualities of the place. But the people who had covered this little spot with their culture symbols were dealing with a very tangible entity anyone could see. Having come face to face with their reality, I realize they were right.

At Cowhead Mesa I have not been privileged to see the *genius loci* in such a vivid way. I've studied the petroglyphs for clues, reflected on the patterns of rodent bones in the owl balls that litter the floor beneath the cliff, tasted the honey from the hive (though honeybees are European and weren't here in the 1750s) that fills a cavity 15 feet up the face, but have never quite touched *it.* The closest I've come is by climbing the mesa and watching sunset from the northwest overlook, a series of shelves sculpted from green sandstone. It's a place for seeing animals—always white-tailed deer, a gray fox once, a bobcat the first time I made the climb—and there are more glyphs here to indicate that you are not the first. And it is a place for exercising the ancient human love, of the old plains primate who has been liberated by bipedalism for far-seeing, for great vistas and a commanding view from above. Perhaps this was the elemental, important thing that was the original *genius loci* of Cowhead Mesa. Now, because a long-dead Comanche let it inspire him to incise on its face the symbols of his world, Cowhead Mesa has become articulate enough to work its magic on our culture, too.

THIS MORNING Tule and I stay on the opposite side of the Double Mountain Fork from Cowhead Mesa, so we aren't tempted to explore it, and follow one of the ubiquitous cattle trails that cut up this rough ranchland. This makes for easy walking, but August in this desert is a quick drain on any life exposed for long, and within a few minutes I miss Tule, then hear him splashing into a pool he's found in the riverbed. Overhead the Texas sky is an unmarred cerulean, and except for the droning of cedar flies and the distinctive falling notes of a canyon wren in the cliffs across the river, the Double Mountain Fork Canyon has gone to sleep against the summer glare.

Tule shows up and shakes a spray off his coat that seems to evaporate before it hits the ground, and with no particular itinerary except to explore what interests us we leave the river and angle off toward the north rim of the canyon, a mile distant. Fifty thousand years ago, when the Double Mountain Fork began to cut down through the surface of the plain, it evidently followed a natural geological break into an ancient lake bed. The result today is a different geology between north and south rims, the south rim composed of a high, sharp Fredericksburg cretaceous chalk that is full of fossils from the old Cretaceous seas. The north rim, though, is capped by Ogallala caliche, which lies directly atop the green shales, the stratified clays, and the knife-edged, bright red sandstone of the Triassic age, a time when dinosaurs stalked and munched our mammal ancestors. It was in this canyon, in fact, that fossil specimens of *Protoavis,* believed now to be the definitive transitional lifeform bridging the hopping reptiles and birds—like the trilling canyon wren I'm hearing—were discovered in 1986.

The north wall of this canyon is a peculiarly compelling stretch of country to me. For the most part it is what one would call, on the Northern Plains, a badland. So friable are the Triassic clays under runoff, so relentless the beat of sun on these south-facing slopes that, except for cacti and yuccas and other desert plants, ephedra and sagebrush mostly, big expanses hundreds of yards in extent appear absolutely barren of vegetation, a multicolored land of clays and sandstones sculpted by water and wind into fantastic Saturnian shapes. Here, erosional necks of hoodoos rise off lumpy red clay mounds and look from a distance like preserved earthen dinosaurs guarding their old haunts. Even the summer sun can't wash out the high color of the Dockum sandstone, which alone seems to hold this canyon wall upright. This is the most southerly appearance of a formation that 150 miles to the north, in the canyons of the Red River, erupts into towering spires, buttes, and fins of gleaming red slickrock.

It's a place that would repel many people, but I find its eeriness a strange draw, and since locating it a few years ago I've repeatedly returned to experience what I call the Technicolor Cliffs under different conditions, crawling around like some sort of inefficient ant in this natural playground. Tule has been here too, knows there is no water, and is not thrilled that we're heading toward the baking badlands at this time of day. But the river is not far and he indulges me. We enter the badlands.

Despite some superficial similarities, the canyons of the Llanos are all distinctive, one from the other. The stamp of the desert distinguishes this one. Popular Texas opinion notwithstanding, according to desert ecologist Forrest Shreve's long-accepted definition, the Southern Plains is not a true desert. At least not yet. It gets too much rainfall, spread over too much of the year. But this canyon, along with the Muchaque Valley 20 miles to the south, comes close.

Ten miles farther downcanyon, where erosion has peeled back the opposing walls several miles and the river coils through an entrenched, secondary canyon, additional mesas many times larger than Cowhead stand 300 feet above the floor of the plain. One of them is Flattop Peak. It has played an important role in botanical investigation in this region through a field of study known as island biogeography. Most of this area was badly overgrazed when Hispanic *pastores* herding sheep and Texas cattle ranchers with their Longhorns overstocked the range, suppressed fires, introduced almost always harmful Old World plants, and unleashed the spread of certain native ones through ecological disturbance. Island biogeography argues that steep-sided mesas like Flattop preserve on their tops relicts of the original vegetative composition. Studying it is one of several accepted techniques for looking back through time to try to see what pre-Euroamerican ecologies were like and for evaluating environmental changes that have taken place in the change from Native American land uses to Euroamerican ones.

In fact, Flattop Peak is almost a textbook example of some of the insights offered by island biogeography, and what it tells us is that there has been a serious and far-reaching change in vegetation and accom-

CLOCKWISE FROM UPPER LEFT:
Cholla cactus; Dotted gayfeathers of a healthy mixed prairie; Hoodoo and eroded mounds of the Double Mountain Fork Canyon

panying animal life in the Double Mountain Fork country and that a noticeable desertification has occurred here over a mere century. Judging from the species composition atop this mesa today, commercial grazing, even in an ecology that has evolved with endemic roaming herds of grazing species, is one of the quickest avenues to ecological simplification and desertification. Not only does a 30 percent greater diversity of species exist on Flattop compared with nearby ranchland, but also the dominants on the mesa top comprise a moderately moist, or mesic, prairie community, that is hardly a page from a desert text—Pinchot juniper, shinnery oak, the dainty little feather dalea shrub, vine ephedra (a world-famous medicinal herb here often called Apache tea), plus a rich mosaic of midheight grasses like side oats grama, blue grama, and rough tridens. But the rolling valley and canyon floor lands that surround Flattop are cattle lands, where comparative range studies show soils six times as compacted as those atop Flattop Peak, with mesquite five times more common, inch-high buffalo grass dominant to the virtual exclusion of the midheight types, at least a full dozen exotic weed species growing, and a plethora of cactus species blanketing the country.

Where I hike nowadays, in the upper canyon, the topography is rougher, the soil conditions and microhabitats more diverse by far, yet many of the same ecological trends are apparent.

Twilight at the base of the Technicolor Cliffs

There is justice but little poetry when ranchers curse the desertification they have helped bring on and try every twentieth-century technological fix, from mass spraying of herbicides to chaining cacti and bulldozing the proliferating junipers and mesquite, to reverse it. Meanwhile the desert relentlessly spreads, making the country less and less useful in economic terms but ever more fascinating as a laboratory of natural history.

Cacti, for example, much as ranchers hate them, are a marvelous metaphor of adaptation, a symbol of the Southwest for many people. Throughout the Llanos canyonlands there are about fourteen native species, actually ancient invaders from the South American dispersal center of Cactaceae evolution, which in a short 20,000 years has sent its fleshy green offspring as far north as the Arctic Circle. Ten of the fourteen grow in this canyon, spiney, ribbed, red-fruited, tubed and padded and barreled desert marvels, shaped for water storage and photosynthesis without leaves. Walking the Double Mountain Fork Canyon no one could miss them; in fact, the trick in many places is making sure you *do* miss them. Unlike mesquite thorns, cactus spines carry no toxins, but most are barbed and they can fester and take weeks to work out.

Save for the candlelike genus *Echinocereus,* the fleshy *Mammillairia,* and the nasty *Echinocactus,* a short, fearsome little brute of a barrel cactus the Hispanics call *manco caballo* (devil's head), all the other cacti of the Double Mountain Fork Canyon are Opuntias, one of the most primitive of cacti genera. The familiar prickly pear is of the genus *Opuntia,* and at least three kinds are found here, including the Comanche prickly pear, a type with pads the size of small frying pans. But my favorites, and the species that strike me as iconic for this canyon, are the chollas. The spindly little desert Christmas cholla, with its thimble-sized, watermelon-flavored fruits, is widespread in all the Llanos canyons. So is *Opuntia davisii,* although it is far rarer, a low, imposingly spined cholla named by a loyal ex-Confederate botanist for Jefferson Davis. Its strategy for replication is self-dismemberment, the joints seeming to leap space to hitch rides on passing animals in an effort to colonize new locations. But the two largest chollas, the 6-foot-tall tree cholla and the similar but slenderer Klein cholla, are found on the Texas side of the Llanos in great numbers only in this canyon. Interspersed among the junipers and the ubiquitous algerita bushes, growing head high like green, many-armed extraterrestrials who, one suspects, bow and mince around one another when no one is there to see, these chollas are the most graphic examples of human-induced change here. Not a single one grows on Flattop Peak.

The sun climbs higher, not quite vertically overhead two months after the summer solstice, but high enough. My shadow gathers about my feet, and the dry heat operates deceptively, evaporating perspiration so quickly that a dehydrating person can be unaware how rapidly precious bodily fluids are being sapped away. There's a slightly unbelievable (if only because movies have made the images such stereotypes) example of the danger from local frontier history, an episode known as the "Staked Plains Horror." With real life desert pathos a white army captain named Nicholas Nolan took his command of black Buffalo Soldiers onto the Llano Estacado during a dry July and August in 1877. Five of them died, the moisture baked out of them as the force ran out of water, then disintegrated in the country just west of here. Dead Negro Draw, just two miles upcanyon, still appears on modern topographical maps as a reminder of the fate of one of them. I follow Tule's advice and join him in the shade of a high bank along the little river, which in this stretch runs thinly but brightly over varicolored gravels.

Two hours pass. The shadows begin to stretch out. Half a mile away, atop the north rimrock, is movement—a head, two, five heads, adorned with lyre-shaped horns, seemingly suspended in space until I realize that at this angle the white underbellies have merged into the white cliff, that the hanging heads belong to the small remnant band of pronghorn that still lives here. Motionless and in deep shade as we are, they are looking right at us with those marvelous plains binocularlike eyes. Pronghorn evolved here, were one of the few native quadrupeds to survive the Pleistocene extinctions; only a century ago there were millions of them in this country. Now they are virtually extirpated on this side of the Llano Estacado. This tiny band is the only one I know of in this canyon, although there are others not far south toward the Muchaque and Colorado River country. I raise my hand in greeting and instantly these are gone, only the dust hanging when the buck's snort reaches my ears. So much for the legendary curiosity of the pronghorn. But who can blame them, considering what the human hand has symbolized? Only cynical pronghorn here now.

WITH LATE AFTERNOON the air begins to cool appreciably, color deepens in the red cliffs, the Double Mountain Fork Canyon begins to come to life. I shoulder the pack, adjust the straps against 38-year-old muscles, and Tule and I start off again. But for some reason, after 30 or so yards, I stop, stand thinking for a minute. Upriver the canyon shallows into a wide V; upriver are overhangs, good overnight places where buffalo hunters and cowboys once made their camps. Upriver also reminds me of J. Wright Mooar, he of fabled white buffalo fame, who killed 4,500 animals here in the autumn of 1876, and even if he did cure the meat and haul it in to Fort Worth, I don't think I want to cogitate on Mooar and the buffalo slaughter tonight. When at length I step out hard toward the north wall at a downcanyon angle, Tule stares for a long minute, then lopes past me.

There are plenty of things to look at now, but one of those I especially watch for is a large, dark-purple, afternoon-blooming flower that is found in deep sands like these along this river. It is the Texas poppy mallow (*Callirhoe scabriscullа*), a federal- and state-listed endangered species known from the upper Colorado in Runnels County but not yet from this canyon, although some ecologists believe it is here. I've seen its relative, *C. involucrata*, the cowboy rose, or wine cup, along several of the stream banks of the more northern canyons. It is out of flower by this late in the summer, however, and probably the endangered type is, too. The Texas Nature Conservancy is very interested in this species, and to find a colony here would be a real boost to establishing a protected preserve in this canyon. August or not, I look hard for it.

A big collared lizard of the high rimrocks

On a gravelly flat near the canyon wall there is quick movement. A buff-spotted ground squirrel, its bushy little tail erect like a flag, zips through the prickly pear like a speeding bullet. As desertification spreads here, more rapidly than its rate of spread in modern East Africa, it brings with it a host of desert creatures that once were only found farther south. Little desert shrews and pygmy mice, Western pipistrelle bats, several species of *Diopodomys*—Ord and bannertail kangaroo rats, marvelous, bounding little water stills that need none from external sources—are all found here now. Most of these are mammals of the wild, uncultivated deserts and are rarely seen because part of their desert repertoire is nocturnality. But not this little torpedo. It's the Mexican ground squirrel (*Citellus mexicanus*), about at the northern extremity of its range in this canyon. I mark where this one is in hopes of returning to photograph it. Mexican ground squirrels have a range of only about 100 yards. Unless a coyote or bobcat catches this one, I should be able to find it another day.

Like most of the arid Southwest, the Llanos canyonlands have a healthy reptile population and a larger amphibian population than one would think. Reptiles in particular seem well suited to desert country because of their ability to adjust their internal temperatures to external conditions. While it narrows the thermal range in which they can operate, this is an adaptive strategy that allows them to reserve energy for optimum conditions.

Not surprisingly, as the evening cools, cool-blooded critters are everywhere. Little whiptail lizards, inspiration to a few of the more radical feminists I know (almost all whiptails are females, and they've figured out a way to reproduce without males being necessary to the task) scurry everywhere across the rocks along the river. Tule chases a skink that slithers for cover when a slab gives way. We're too low here to run into any of the big, brightly marked collared lizards that live up in the rimrocks. But we do see a few Texas horned lizards and one quarter-size round-tailed horned lizard, a species of the deep southwestern deserts that is not very numerous here. The numbers of both species are diminishing in West Texas, evidently because of pesticide accumulation from the insects they eat. The amount of Roundup, Asana, Spike, and other pesticides applied on the High Plains of Texas is staggering. But the impact on biological life in the canyons below is mostly unknown. It's a good bet that the inoffensive little horned lizards aren't the only innocent bystanders.

Despite Tule's misgivings, I have a destination in mind. Well before sunset we cross the oil-field road, have it two miles behind us when the sun begins to gather into an orange ball for its dive into the High Plain. No clouds and no postcard western sunset tonight. But the nighthawks dip their erratic, booming flight overhead and the red in the cliffs is graying, the shadows blue twilight with dark beginning to cup the mesa tops, so I call Tule away from the darting little desert cottontails.

We make a dry camp in a sandy cove at the base of one of the most imposing cliffs in this canyon. Half our water is gone but there is enough for coffee and for the dog, and the truck is only five miles away. And this is an interesting place. Above us the wall is a thinly vegetated erosional slope for about 160 feet before abruptly effecting transformation into a 100-foot vertical sandstone cliff that extends for a quarter mile downcanyon. Raptors and owls nest in this cliff and exotic bones and chunks of petrified wood lay scattered on the slope. But the canyon spreads out like a bay here, the opposite rim nearly six miles distant across choppy mesquite and cactus country.

I make a juniper fire. Within a few minutes it is popping the way juniper does when it releases its energy, lighting and perfuming the little cove with a scent that momentarily puts me in smoky Zuni villages in cold New Mexico winter. The planets are out by the time I rake coals aside for the coffee, and a few thousand stars glitter down at the first steaming sip. As the coals die to glowing, something big whooshes over. Adolescent coyotes, straining and sounding ridiculous, start up, triggering a ten-minute coyote chorus from several different family packs. Tule perks his ears but utters not a sound. I finish my coffee and crawl in the bag, worm around a little to set the sand, and lie staring at the stars.

PEOPLE WITH DREAMS have looked at the Double Mountain Fork–Muchaque country and seen several versions of paradise. For the original hunters, it must have been the clear, gypsum-free water of the canyon that drew them. The Archaic and Neo-Indian cultures, including the famous Garza hunters and the pottery-making Palo Duro culture, which probably sprang from the most easterly Jornada bands of deserts farther west, seem to have found life especially satisfying here. Fire pits where they processed cactus and mortar holes where they ground the seeds they gathered appear frequently in the sandstone shelves along the river, and bones from the bison they drove off the cliffs still wash out in the rains. In the north rim of the Muchaque are overhang caves, littered with the remnants of woven materials, whose ceilings are covered with pictograph handprints that deliver a time capsule message: "We were here, and we know some day you will come and wonder about us."

The Apaches and Kiowas saw different possibilities here. Excepting the early prehorse Apaches, who either arrived with or adopted a dog-and-travois culture, these were horse people who focused on the enormous biomass of bison, a cultural vision that made larger human populations possible but intensified the importance of available water. And horses and bison products meant trade with caravans from the pueblos and plazas on the Rio Grande, for both culturally and politically the Southern Plains canyonlands entered history as a part of El Territorio de Nuevo Méjico.

Scholars who puzzle over such things have recently argued, on the basis of botanical evidence, that the great plains ravines, or *barrancas,* where Coronado's force camped could not have been as far north as Palo Duro and Tule canyons but were somewhere here, on the southern Llano Estacado. Spanish records from Santa Fe begin mentioning "Muchaque" by name as far back as the early 1700s. According to one of the last Comanche bands to camp in the Muchaque Valley (in 1876), the name itself meant "looking for the traders to come and bring tobacco." Hence today's Tobacco Creek, the main headwater stem of the Colorado River, and Muchaque Peak, presently dominating a valley of pump jacks dedicated to feeding our culture's insatiable thirst for fossil fuels.

Then there was the cereal king, Charles William Post, at the vanguard of a group come from wetter, lusher lands, who thought he would simply remake the region to create paradise and in 1906 founded a famous Texas colony ten miles north of the main Double Mountain Fork on juniper uplands just below the Caprock.

Post's dreams have largely been forgotten on the late-twentieth-century Llanos. It was a nineteenth-century dream anyway, a utopian vision that on the face of it would seem to have had more in common with Brigham Young's Mormons or with the Greeley Colony in Colorado than with the West Texas that was emerging by the time of Post's death. But his was a western colony that touted "individualism" (of the sort, presumably, that had made Post's fortune), and that placed faith on the American certainty that money and technology could take just about any country and turn it into whatever one wished. Even to remaking the weather.

Post's life offers an example of a man driven to achievement because of misery. To the age of thirty he had had a nondescript career as a traveling salesman of farm implements. But the plow and hoe business lost its romance when he was struck with a stomach ailment, most likely an ulcerating tumor, in 1884. The pain drove him to West Texas to cowboy in the high, dry climate that doctors prescribed for a variety of ailments in the last century. No relief. Relocating in Battle Creek, Michigan, Post tried dietetics, hypnotism. Nothing worked until he began to experiment with decaffeinated coffee and a variety of grain cereals as breakfast foods. The rest is the history you've already anticipated. Postum beverage mix was the product that made his empire. The cereals, Post Toasties, Grape Nuts, and one called Elijah's Manna that didn't do quite so well were its descendants.

He became the quintessence of the nineteenth-century capitalist, despising labor unions and socialists, thorns in his side almost as galling as the stomach pain he disguised but that continued to haunt him. The West Texas colony was designed to prove all the Progressive Era social-engineering "rainbow chasers" wrong. In 1906 he handed nearly a million dollars to a group of bewildered small-time ranchers and got title to 225,000 acres in the Double Mountain Fork country. "We will soon begin to make a showing in the wild and woolly West," his foreman wrote. Enlightened despotism, American style, had come to the Southern Plains.

Post City was indeed a monument to individualism. One man planned and financed the entire town, right down to the kind of wallpaper used in the residences. By 1914, 300 houses, stores, and public buildings were ready for settlers to move into. There was gravity-powered running water in the houses, and each 160-acre plot had 3 acres planted to an orchard before the colonists arrived. The Postex cotton mill and textile factory were built. Humid-land trees of all kinds were planted. Apparently Post had a rather peculiar concept of "individualism." Sod-house settlers from the Northern Plains would have thought they had gone to heaven had they been set down in early Post City.

There was only one hitch. It didn't rain nearly enough to suit Post. Evaporation rates were astounding. Early experiments with deep-level water wells proved that there was plenty of water underneath the High Plain itself, but to be centrally located as the county seat Post City had been built in the escarpment breaks below, where the well water was too gypy to drink. Wind-powered pumps sent water from the plain into a reservoir to water the town, but to Post that seemed makeshift. Hell, he had founded a business empire, and this was America, after all. Why not just take nature on head-to-head? Why not make it rain more?

The southwestern Indians hold ceremonies and dance and pray to the gods for rain when the droughts hit. The fatalistic New Mexicans may gather in their plazas to talk about drought but accept their lot with a shrug—"What can one do?" In a typically American response, Post tried to blast moisture from the grasp of the gods with dynamite. The Texas Baptist, in the universal southwestern tale illustrating the divergent relationships with nature held by Indians, Hispanics, and Anglos, is supposed to have remarked: "The Lord will look down and say, 'Look at those poor ignorant people. I gave them the clouds, the airplanes and the silver iodide, and they didn't have the sense to put them together.'" It's a sentiment that still fuels rainmaking efforts on the plains.

It was a widespread folk wisdom in Post's day, seemingly vindicated by certain Civil War engagements, that explosive concussions could cause rain. Post wasn't the only rainmaker in the West to try it, but he had both money and backing from Du Pont Powder Company to try it on an unprecedented scale.

They were called "rain battles," and between 1911 and 1913 Post's engineers held twenty-one of them, setting off an average of 300 pounds of dynamite per battle. Kites were used at first to hoist the payloads aloft, but when that proved extraordinarily hazardous to the kite flyers, the charges were set off from stations along the Caprock edge.

How well did it work? It worked at about the same success ratio as the Pueblo dances. Even though Post stacked the odds by staging the battles only on days when rain seemed likely, out of eighteen battles for which records survive, only three produced appreciable rain.

None of which mattered to Post. But, although he did not lose faith in his rainmaking ability (dynamite for fifteen more battles was laid in for 1914), he did lose faith in himself. His cereals and his decaffeinated coffee had made him rich but they hadn't made him what he really wanted to be: well. In agony with stomach pain, he sent a bullet crashing through his brain in 1914.

His monument to utopian individualism is just another West Texas town today. But his commitment to remaking the Llanos country into a humid oasis of the West has run like an underground river through twentieth-century plains history.

SOMETHING has intruded on my sleep, and for a long, groggy moment I cannot tune in the sense that sent the message. I sit up, awake enough to notice that Tule is also up, and I focus along the direction of his muzzle.

Too dim to see despite the orange sliver of moon in the east, I hear the faint yet certain sound of footfalls, the slightest snick of sliding shale. An almost imperceptible movement of air stirs the hairs on my arms, and a few seconds later there is a violent exhaling snort and then the fading clatter of hooves on rock. Tule starts but sits back down at my voice. A deer, and as the clatter lacked a pogoing rhythm, probably a whitetail.

There are no intimations yet of daybreak, and the cliff behind cuts off my view of the Big Dipper, but the waning moon has already risen. Since dawn twilight is such a magical time to travel the desert, I slip free of the bag and start to work on the fire. Within minutes its bright licking flames have cut the night chill. I brew a quick cup, give Tule a long drink and a half a granola bar, enjoy my last peach remembering John Muir's famous description of eating grapes in the desert, the stored juices under drum-tight skins bursting in his mouth and causing the "finest nerve waves imaginable." From long habit I zing the peach seed at a distant rock. Fresh and excited in the cool twilight, Tule is on it in a flash. The fire pit covered with rocks and sand, the bag put away, we set off through the dawn.

THE ONE HUMAN VISION that hasn't disappeared in this country is the one held by those who herded animals. The New Mexican *pastores,* among them the Garza family for whom this county is named (although typically they are not even mentioned in the county history),

Moonrise and red cliffs

herded many thousands of sheep here by the 1870s, and in the grassy upper canyons and draws of all this Llanos canyonlands country, plus the Canadian Gorge and Breaks. New Mexican sheep had already cropped the Southern Plains grasslands for nearly a decade when Celtic cattle ranchers arrived to claim the country for Texas. While the Indian wars are famous, no one has paid much attention to the war between New Mexicans and Texans over the Llanos. But there was one, and the sheepherders lost it. Fabiola Cabeza de Baca, whose father and friends participated in this Hispanic advance and retreat, describes it well in her memoir, *We Fed Them Cactus.*

The cattle ranchers won the prize, and although up on top of the plain their empire gave way within half a century to crop agriculture, for the most part they still have the canyons, which were always the core of their ranches. Charles Goodnight created the mold when he drove his herd down from Colorado into Palo Duro Canyon in 1875.

Canyon walls made fencing unnecessary, and all one had to do to keep settlers out was file claim on all the water; waterless lands in between could be ignored since they were impossible to homestead. Losses to predators were intolerable, so hire poison and bullet vendors to take care of the inconvenience of wild animals whose only remaining prey was cows. Impose various puritanical rules on the cowboys and squash any dangerous intimations of unions among them. Then stock the range for all it was worth and sit back and watch it make money, up to 200 percent a year in the first few years.

Thus were the Southern Plains canyonlands settled between the 1870s and the 1890s. The early ranch names are mythic by now: Goodnight's JA in Palo Duro, Tule, and Mulberry Canyons (1876); the Rocking Chair between the Salt and North forks of the Red (1883); the Frying Pan (1881), which developed Amarillo; the Bell Ranch (1870) on choice parts of the Canadian Breaks below the Gorge; the Two-Buckle (1882) and McNeil (1883) in Blanco Canyon; the Kidwell Brothers (1881) and the Western Land and Livestock Company (1884) in Yellow House Canyon. In this part of the canyonlands the Slaughters, C. C. and John, founded the most famous of the early ranches, the U Lazy S Ranch partially in the Muchaque and the Square B and Compass in the Double Mountain Fork Canyon, both started in the early 1880s.

The amazing thing is not that the bulk of the canyons are still in the hands of the ranchers, but. that in many instances they are still in the hands of the same families that pioneered them, or have changed hands at most only two or three times. Of the three major ranches in the Double Mountain Fork Canyon, the Macy and U Lazy S ranches are still owned by John Slaughter's descendants, and the Miller Ranch dates from the early 1900s. This kind of familial continuum deteriorates in the canyons that lie near Amarillo and Lubbock, but not by much. My little piece of Yellow House had changed hands only five times when I bought it almost exactly one hundred years after the founding of the Kidwell Ranch.

What one has to grasp about ranching as an ecological adaptation is that it is, first and foremost, a modern business made possible by the Industrial Revolution and that, as a business subject to the cycles of the market, it is driven by the same economic margins that affect Wall Street. Ranching land stays wilder than farming land, but only because ranching is both land and labor extensive and because—usually anyway—success depends on retaining the native vegetation intact instead of replacing it wholesale. Entice a few hunters to pay big money for hunting privileges, and the wilderness incentive clicks several notches higher.

THE EFFECT of light on the plains is wonderfully variable. In the shadowless glare of summer midday any country within forty degrees of the equator flattens and dulls. Interest, romance, myth disappear. But slanting light with its shadows and crisp color exaggerates everything upright. Rock slabs stand boldly off the ground, foot-high silver sagebrush casts tree shadows, a cholla cactus towers in prickly outline over the desert floor like a lightning-struck pine. The sun-cracked clay floors of rain pools take on texture, and every little arroyo is alive with the potential for adventure.

I am kneeling by a meandering set of tracks in the magical world of sunrise giantism when I see the bobcat. Not as big as a Smilodon, and even the quick backlit image as it bounds, feline fashion, behind an algerita bush can't convert it into a cougar. But there is that one instant, before the synapses relay a recognition code.

Thing is, it could have been a cougar. The canyonlands were once the center of their range on the Llanos, before cowboys and government hunters placed at the disposal of ranchers eradicated some 360 of them in West Texas in the great twentieth-century predator war. Small remnant populations still exist in the Canadian Gorge and in the Red River canyons. And here. Rancher Riley Miller killed one in this canyon a few years ago, a big female.

Tule steps out of a cholla patch 100 yards down the arroyo and shakes himself, and we start again, Cowhead Mesa upriver less than two miles away. The sun is warm on my shoulders as blue morning dissipates. A male curve-billed thrasher glares through red reptilian eyes from a juniper. The cat-algerita poetry playing over and over in my head, I remind myself to send Riley a copy of D. H. Lawrence's "Mountain Lion":

> *So, she will never leap up that way again, with*
> *the yellow flash of a mountain lion's long shoot!*
> *And her bright, striped frost-face will never watch any*
> *more, out of the shadow of the cave in the blood-*
> *orange rock.*

THE RANCHERS, a few of them anyway, have come a long way in their thinking, of course, a long way since Goodnight hired Gus Hartman and Rufe LeFors to poison all the bears and coyotes and wolves and cougars in Tule Canyon, a long way since Rollie Burns found an eagle roost in the Double Mountain Fork and had his cowboys go in at night and massacre the lot of them. Biologist David Brown blames their southern heritage for the early ranchers' and cowboys' almost fanatical hatred of wild animals, especially predators, but it wasn't that simple. The war against the wild was the war for civilization. It made no difference whether one was a southerner.

At any rate, they killed everything, not only all the predators but also all the deer, all the antelope. They killed off the turkeys because they were easy, the prairie dogs because they ate grass that a cow might want. What animals were left when the Great Depression came were polished off by pot hunters. Then, when it was all gone and there was nothing but the land left, and even that changing before their eyes, the ranchers started trying to put it back.

The early Southern Plains ranchers—there were exceptions—didn't seem to want to know the natural history or ecology of the land. One of the famous quotes in ranching history came from a meeting of West Texans in 1898, who adopted this statement of their interest in nature:

"Resolved, That none of us know, or care to know, anything about grasses, native or otherwise, outside the fact that for the present there are lots of them, the best on record, and we are after getting the most out of them while they last." The ones who have survived the intervening century have learned the folly of that attitude.

Riley Miller and his son, Ben, whose adjoining ranches cover most of the lower canyon, are exceptional but illustrate the direction. Riley is enamored with the wilderness West, maybe because he's a hunter, and although you have to watch where you step when he's really going, he's dead serious when he says (as he does a lot), "You gotta leave it better than you found it." For three generations the Millers have been trying to patch the wild back together in the Double Mountain Fork. "There was nothing here in the fifties but a few cottontails, and they was scared to death," Riley says. But his father reintroduced wild turkeys, and whitetails set loose on the Miller and U Lazy S ranches began to make a comeback. Desert mule deer, once native to the Llanos canyonlands, were stocked in 1957, although the current spread of brush seems to favor the whitetails. Riley brought in bison from South Dakota (there are about sixty-five now), elk from Wyoming. On the ecological downside, he and Ben have also successfully introduced a few exotics, aoudads among them although they much prefer a desert bighorn-moufflon cross that seems to thrive even better in those tall, chalky rimrocks of the south canyon.

Like a lot of ranchers, especially in Texas, the Millers have come to the realization not only that the wild is more fun to be around than a country with nothing in it except vapid, screw-loose cattle but also that the wild plus recreational tourism can make you plenty of cash.

IT IS HOT AGAIN when we get back to the truck, and although it is too dry to sweat, there is a certain relaxed feeling in my neck muscles, a slight difficulty with quick focusing that signal the onset of dehydration. Tule has been wallowing in the river for the last mile. Out of water since daybreak, my own cells are shriveling. But I long ago learned how to reward myself after hiking in desert canyons. After a couple of long pulls on the canteen in the truck, I open the ice chest and then sit down against a front wheel to savor the sharp, cool bite of a Tecate.

From across the river, from the near rim of Cowhead Mesa, it seems, comes a mournful cooo . . . cooo . . . cooo. Then silence. Then again the cooing, until it is drowned by a pair of jets, pilots from the base west of Lubbock, testing whether their stuff is right, I guess.

I swill down half the beer. The roadrunner up there, doing his ground cuckoo number and watching me—he knows. He has a salt gland in his beak for hot weather, he can lower his body temperature up to seven degrees to conserve energy, and when the cold wind sweeps the flats he can warm himself by exposing a dark patch of skin on his back that acts as a solar conductor. The old-time New Mexicans used to call him *paisano*—countryman. He is the product of adaptation to the Southwest since the Pleistocene, and the creep of the desert here suits him just fine. He knows he'll stick.

I finish my beer, and Tule and I climb into the truck.

OPPOSITE PAGE:
Queen of the canyonlands, plains cottonwood in autumn splendor

Chapter 3

Grassy Gorges of the Brazos

The catalyst that converts an environment into a place is the process of experiencing it deeply—not as a thing but as a living organism.
—Rene Dubos,
The Wooing of Earth

Once in his life a man ought to concentrate his mind upon the remembered earth, I believe. He ought to give himself up to a particular landscape in his experience, to look at it from as many angles as he can, to wonder about it, to dwell upon it. He ought to imagine that he touches it with his hands at every season and listens to the sounds that are made upon it. He ought to imagine the creatures that are there and all the faintest motions in the wind. He ought to recollect the glare of noon and all the colors of the dawn and dusk.
—N. Scott Momaday,
The Remembered Earth

Here I go rowin
Thru Lake Innifree
Looking for Nirvana
Inside me.
—Jack Kerouac,
Mexico City Blues

I AM FROM THAT GENERATION. You've seen us. We're all over the West, from Montana to Texas, and in certain rural areas in the East and Midwest. We grew up reading *Mother Earth News* and *Whole Earth* catalogs and various other eclectic back-to-the-land rags, and we share a certain homesteader vision that is at the core of the brief American experience. Some regard us as the enemy, because our insistence that we live in natural settings has fueled the division of large private landholdings where a mild form of wilderness seemed protected because no one but the owners could get to it. The poets of the movement, Gary Snyder, Edward Abbey, tell us we're the heroes of something called bioregionalism and the reinhabitation of the American landscape. Like most humans involved with the natural world—Indians, mountain men, landscape artists—we're a little bit of both. But we are here; we're a fact of late-twentieth-century rural life.

The extent and flair of the modern back-to-the-land movement are exceedingly variable. Daydream about it and the bottom line glares back. How much money you got? How much land can you buy, and where? You want a solar-heated adobe tucked away in the junipers south of Santa Fe or a horse ranch in northwest Florida, best you be a very successful yuppie. Hammering together your own place on a small piece of land in Montana or Colorado doesn't take as much. And on the Southern Plains, where farm- and small-town life makes most folks dream city rather than country dreams, building a cabin on 10 or 15 acres is hippie homesteading if you can find the spot.

In the early 1980s I found a spot.

SOUTHEAST and northeast of Lubbock, two beautiful, rock-walled canyons slice grassy valleys 200 to 300 feet below the surface of the Llano Estacado. Blanco Canyon, whose name dates back to Comanchero times despite its rendering in West Texanese ("Blank-o"), is the deeper and longer, stretching 22 miles out of Running Water Draw down to a wide, eight-mile mouth where the White River and its tributary springs cascade waterfalls off Triassic shelves. Yellow House Canyon is its twin on a 30 percent reduced scale, a vertically sided canyon that steadily widens down the 16 miles from Buffalo Springs Lake to Courthouse Mountain, where its walls peel away north and south to become the Caprock Escarpment.

These two prairie canyons are the apotheoses of a canyon type on the plains. Where the Llanos scarp is abutted on the east by rolling lands no more than 300 feet lower than the surface of the High Plains, the canyons are gradual affairs, shallow and long rather than the plunging gorges that occur where the escarpment is hundreds of feet high. Lateral recession of the walls in this kind of canyon has occurred faster than down cutting. They differ from draws in having eroded sharply enough to fashion rock cliffs, but their low gradient streams tend to skate over the top of the Triassic and enter the harder rocks of older geologic formations only at their mouths, where their streams etch entrenched, secondary canyons through the sandstone. Quitaque Canyon is such a canyon and so is the little three-mile-long canyon fashioned by McClellan Creek. The few canyons that come off the Mescalero Escarpment of the Llano Estacado—Alamosa Canyon is one—are of the same type. But none is as compellingly lovely as nature works of the American West as the twins at the head of the Brazos.

Conceptualize them as corridors of South Dakota or Wyoming prairie tucked into an ocean of surrounding Texas cotton-field flats. In fact, they preserve the only native prairie grasslands on this part of the Llanos. Lubbock County, for example, has had 97 percent—that's no typo—of its native vegetation removed. All the 3 percent prairie remaining is in Yellow House Canyon. All the buffalo-grass carpeting, all the winter-orange bluestem, all the blue grama with its ripe seed heads looking like little musical notes, all the side oats grama maturing like a row of feathers hanging from a lance.

So visualize these prairie canyons: opposing, friable rims of gray, pink, or white rock; a dotting of junipers and mesquites setting off walls and mound-shaped swells of wheat-colored grasslands; galleries of cottonwoods and willows and, true, introduced elms and salt-cedars threading the bottoms of the valleys. Standing detached from the walls are elegant, graceful mesas and buttes, their sides sweeping in unbroken line up from the canyon floor, their tabletop peaks exactly on the level of the plain above. There are places, in Yellow House, in Blanco particularly, where I have stood on the rim, or a mesa top, and in my mind raced golden autumn light bareback across and into country so rounded and feminine and smooth that I experienced profound sensual arousal.

THE PLACE was pretty raw, a little bit over 12 acres of snakeweed and mesquite-infested prairie on the south rim of Yellow House Canyon. There was a tar-papered house, framed and enclosed but basically a do-it-yourselfer, a shack of a barn built around discarded power poles, and the ugliest sucker-rod corral ever conceived. There was a maze of electric fencing and several thousand pounds of rusted stoves and other junk—including a 16-foot boiler—on the slope below the rimrock, the result of heroic efforts by residents of nearby towns to fill in the canyon. The homestead had been started from ranchland a couple of years before by a local musician and Texas Instruments employee and his artist soon-not-to-be wife, just about the time both T.I. and their marriage did twin belly rolls. He moved to Dallas and she moved to Austin and they put the place on the market.

Plains centipede

It was raw. My ex-wife visited the first summer, on her way from Taos to Austin. During the night, with a flurry of sliding feet, something tumbled out of the insulation and plopped onto the bed within inches of her face. It was a six-inch centipede and she hasn't visited me since. But despite all, the spot is magical. It sits in a natural cul-de-sac in the valley, the canyon wall forming an amphitheater behind, a stunningly drawn mesa dominating the whole eastern sky. However feasible or satisfying to my instincts it would have been to have wind or solar power, the cabin was in fact already on the electric grid. Once finished out, it would heat easily with a single box-design wood stove. Critically, it had a well and good, cold, slightly mineralized water. In sum, after I chased out a sparrow hawk that was living upstairs, I bought it and it proceeded to take possession of me.

Blue grama in seed

Years later the tendrils have reached out of the caliche-packed soil and coiled around my ankles. I have been changed by the place, and I have changed it. I took down the fencing, hauled away the junk, planted native trees, burned to restore the native prairie. When I bought the place I knew as much about carpentry as I understand about quantum physics. When a good friend who helped me get started told me the story of the fellow who set the vertical line for his stovepipe by putting a level along the barrel of his .410 and firing it up through the ceiling and roof, I didn't guffaw. I was trying to remember which friend had a .410 I could borrow. But I've learned, and have discovered what my grandfathers and a lot of my contemporaries long ago knew, that carpentry has an odd satisfaction about it and that I like it a lot more than plumbing and sheetrocking. I've also discovered something a little more profound. Building a house yourself not only teaches some desirable, probably indispensable, skills about how built environments work, it also becomes an apt metaphor for the human condition. At least my human condition. Growth and change come in spurts and flurries, with long spans of quiescence between. And growth, like putting a place together, is somehow tied to dissatisfaction.

Indian blankets

I have been on this little piece of ground long enough to have learned the plants and the birds and the turn of the seasons in this canyon at a deeper level than for any country I've known, even the woodlands where I grew up. It is absolutely because I live on it, rather than visit it merely. It's given me a kernel of insight into the kind of spiritual ties that tend to develop between tribal people and local landscapes, between any people and a place they occupy for long years or generations when there is a conscious effort to maintain it and understand it as natural landscape rather than one made over with introduced plants and animals. And despite a big chunk of my life spent roaming the outback, living here I have come to know the sky on a new level—the

Blanco Canyon

LEFT: *Yellow House Canyon at its autumn best*

One-seed juniper and Yellow House overlook

cycles of the planets and constellations, the phases of the moon, the movement of the sun across the horizons from solstice to solstice—as I have watched the days and seasons pass over this little canyon hidden away in the stupendous Great Plains of the American West.

ON CERTAIN West Texas dawns, if I strain only a little, I can hear hoofbeats, the creak of packs, and the sound of travois poles bouncing over rocks on the rimrock above the house. Ghosts of mountain men and Comanchero caravans and Kiowa war parties still roam this canyon. Through an act of Romantic imagination, on certain mornings I can conjure them.

They are a part not only of the real history of these canyons, but of the mythic, as well. They play heavy symbolic roles in plains history, and the facts as professional history lays them out be damned. No historian will ever make the slightest dent in folk history on the Southern Plains by trying to point out that Quanah Parker was just one among many capable young Comanche warriors during the Indian wars, that he was not the Comanche leader at the Battle of Palo Duro Canyon (his band was not even present), that he could not have climbed a certain mesa in Blanco Canyon to make the decision to take the Kwahadis into Fort Sill, since the Kwahadis were camped almost 100 miles from there when Isatai, not Quanah, reached that decision. Folk history has made its judgments and does not care if much of the history it honors is myth.

Mythic history has a critical advantage over professional history in winning the hearts and minds of the residents. It's landscape associative. Folk traditions and oral history always have been. Professional history tends to regard a fascination with place as antiquarian. But mythology is all about place. Mythology makes ordinary places the scenes of great events, thus giving them extraordinary power.

Much of the mythic association on the Southern Plains resides in superannuated form in a simple thing we take for granted: names on the land. I wish the West Texas pioneers, like those in the Far West, the Deep South, even New England, had learned and retained more of the Native American names once attached to these canyons. Quitaque and Muchaque have endured, but who today remembers the White River as Tosahhonovit, Courthouse Mountain as Wah-we-ohr (Blowing Mountain), or the stream threading Yellow House as Tah-tem-a-reie (Trader's Creek)? Of course a good many of the New Mexican names are yet in place, among them the essentials of landform. Thus, detached mountains are mesas and dry streambeds are (sometimes) arroyos, rather than buttes and coulees, as on the French-explored Northern Plains. Cañon del Rescate (Rescue or Ransom Canyon), the Hispanic name for the canyon section of Yellow House during most of the nineteenth century, has been forgotten, although a ritzy subdivision in the canyon has preserved its translated form. The final overlay of Celtic is apparent in a place like Yellow House, where we have a Long Hollow and a Pig Squeal Spring. And cowboys and surveyors renamed the stream the North Fork of the Double Mountain Fork and translated the Hispanic name for a mesa (Casa Amarilla) far up one of the headwater draws into the name of the canyon.

I suspect that the mythic and historic associations that give Yellow House powerful meaning to me may not be the ones that inspire most West Texans, who seem to prefer cowboy-rancher and Indian-wars mythology. At least the folk and local histories mention none of the events or people I celebrate.

Like Old Bill Williams, the infamous Taos mountain man, long-haired, bearded, the quintessence of the nature sensualist, riding his buckskin "Santyfee" at the head of a forty-five-man party of trappers (it included Albert Pike, who kept a journal) down the Cañon del Rescate in September of 1832, looking for beaver to trap on the heads of the Brazos and Red Rivers. Williams was all over the canyon, climbing the mesas, shooting antelope. But for those who have the idea of making him a sort of Yellow House patron saint, it might be pointed out that Old Bill was no St. Francis. True to form, he could barely be restrained from bushwhacking a Comanche girl hauling water, and when the party bartered for a tipi, none of them, Williams included, seemed to know how to set it up.

Nor does anyone remember a free spirit like Ysambanbi (Handsome Wolf). Francisco Amangual found him in Yellow House in 1808, leading a band of marvelously fashion-conscious Comanches decked out in the latest in three-cornered hats, long red coats with blue collars and cuffs and white buttons, the effect set off with red neckties. A decade later Ysambanbi's two sons, Ysaconoco and Tanquesji, were still using Yellow House Canyon as their main camp and insisting that it proffered as many advantages as Santa Fe if Spanish authorities wished to conduct a big talk.

The Comanchero traders came here for centuries, kept coming right down to 1891, when barbed wire finally discouraged them. The name "Rescate" came from them, perhaps because Yellow House was the scene of a once-famous event on the Spanish frontier, a result of the Comanche kidnapping in Chihuahua of the young daughter of a prominent Spanish official. When a large ransom was offered for her in a remote *barranca* deep in the Provincias Internas, she sent word to her father that she "had become reconciled" to the Indian way of life.

Or maybe the canyon was called Rescate because that word gradually came to mean "trade fair" for the plains tribes. But mythic imagery stands forth here, too. A young Hispanic boy, sole survivor of a trading party to the Rescate, stands before New Mexican officials with a horrible story. They had met and traded with the Comanches in the canyon. The last night there was a fandango. Liquor was passed around, a stupid mistake. Exhausted, the boy had climbed the mesa over the riotous camp and fallen asleep. In the dawn he awoke to screams. His party, his relatives, were being massacred in the valley below by the Indians with whom they had traded.

Outside my window Turkey Mesa stands silently over the valley of the Yellow House. If it has secrets, it's not telling.

SEASONS do not come suddenly on the Southern Plains. Spring, for example, is a long time aborning. In Yellow House it actually begins in February, when the cool-season grasses begin to green up and the little snakeweed bushes start putting on leaves at ground level, a green lace that imperceptibly climbs the dry stems through late winter. By early March, when the sand sage begins to acquire its own ascending lace, the geese and sandhill cranes commence honking and wheeling their way up the valley. North, seemingly, is a hazy destination for those first few flocks, especially the impressionable cranes, whose leaders are inclined to lose confidence in themselves and abdicate control. Directional anarchy is the result.

The Dust Bowl has left an indelible picture of the Great Plains as dust storm country. It is not a false impression, for any year the winter is dry the spring winds whip up curtains of dust that roll across the countryside like immense brown tumbleweeds, penetrate the tightest window and door seals, cake one's hair and teeth with grit. In three of my first five years in Yellow House, rainfall was far over the average of 18 inches a year; in two of those years there were no dust storms at all. But there were dry years, too, so that a dozen times or more a yellow veil entirely obscured Turkey Mesa, 200 yards from my window. Such are the spring delights of life in the agricultural empire of the Llano Estacado. And while Texans eat dirt and pesticide residues, just across the state line in New Mexico, where the grass cover is still intact, the sky stays blue even with whistling March winds.

The high southwesterly winds and dust do not extend much beyond the vernal equinox, at least not during the wetter phases of the 20-year wet-dry cycles on the plains. In their wake remarkable changes begin in the canyons.

Splashes of color begin to appear in the drab brown. Invariably, a tiny yellow flower, the bladderpod, is the first noticeable bloomer in Yellow House, carpeting the gentler lower slopes with a saffron that makes me think of the canyon as a half-made bed. Within days the bed is finished as violet-purple locoweeds carry color up the higher slopes.

In the towns atop the Llano Estacado the exotic trees from colder climates all over the world are already blooming by the middle of March, and the Siberian elms and tamarisk bushes that are slowly taking over the valley bottom from the native canyon trees do the same, the pale green new leaf of the elms and the faint magenta of tamarisk buds preceding all the natives. Skunk-bush sumacs are the first of the natives to bloom, sending forth sweet-smelling, yellowish blossoms in late March, just slightly before the algerita bushes put out their own tiny flowers. But March and April are tricky times for the natives, with soggy spring snowstorms that feel more like Wyoming than Texas always a possibility. One morning the end of March in 1987 the predawn air was clear and thin and windy and my outside gauge stood at just 11°F. Yet only twelve days later the cottonwood buds popped, and in two days of eighty-degree sunshine clusters of leaves unfurled like arthritic hands. Less able to overcome late freezes, the hackberries and willows wait a couple weeks longer before cautiously opening their own swollen buds. And last of all, usually not until almost May, come the mesquites, semitropical in their origins, their leafing a sure indication, say the old-timers, that winter is gone for good.

Gone, too, by the end of this six-week transition are the vast flocks of western meadowlarks that overwinter here, and more marsh hawks—

short-winged, long-tailed accipiters whose acrobatic hunts enliven my winters—leave every day. But there are compensations. A poor-will that likes the hackberry thicket at the base of the mesa, and whose cry redefines dusk in the canyon, returns in April. Cardinals whistle in the warm days, mourning doves coo, and bobwhites wax lyrical in every direction. The coyotes, whose yapping and yodeling had punctuated clear winter nights during their search for mates, have fallen silent now that den sites are being prepared. But a pair of great horned owls that nest in the cliffs behind the house is back, and my summer compadres, the little lark sparrows, show by early April. It's a time when birders can see four or five dozen species a day, especially near water. I see less diversity on my place, but it's worth it to watch carefully. Dazzlers like painted buntings or golden eagles, or, just once, a pair of white whooping cranes flying side-by-side in a flock of about 300 sandhills, appear often enough to enrich the spring migration.

Migrating sandhill cranes, icon of the equinox skies on the Southern Plains

By May's beginning the standard prairie complex of wild flowers is laying down frenetic color in the canyons. If there has been moisture. Indian blankets, Tahoka daisies, plains blackfoot daisies, and sleepy daisies line most of the roads through Yellow House and Blanco, and yellow plains zinnias speckle the sides of the drainage draws. Needle-and-thread grasses wave like little flags in the breeze. (Pull out the ripe floret and you'll see why its common name is so appropriate.) The blooming of the local *Yucca angustifolia* is the crowning accomplishment of spring in the canyonlands. When it is over, and the little white yucca moths have performed their pollinating function in one of the West's finest examples of plant-insect symbiosis and coevolution, the drama ends under a frenzy of insect demolition that tears the creamy flower clusters to shreds.

Although it may seem so to those who live in the city and visit the canyon only to drive through and sight-see, summer is no more static a season than spring. Its rounds are just more subtle, easier to miss.

And because of these canyons' remarkable ability to respond to the varying moisture patterns of each summer differently, the summer progression is never quite the same from one summer to the next. Generally, I mark the cycling of the summer by a handful of plants that dominate my landscape: the yucca miracle in May, the blooming of lavender-flowered beebalms in early June, the profusion of waist-high basket flowers around and after summer solstice, their replacement by equally tall and numerous curlycup gumweeds in late summer, by which time the transition to autumn is underway, the grasses yellowing and setting seed and stiff gayfeather spikes beginning to show purple in the grass.

Problem is it doesn't happen this way every summer, and reading Joseph Wood Krutch and Edwin Way Teale, I hope they are aware that the seasons they describe so well are to some degree singular events. On the plains, where almost every patch of vegetation is in a state of succession because of burning (or the lack of it) and where moisture (or lack of it) programs variations from year to year, no two summers are really alike.

As I write this, the drought of '88 has produced a summer like no other I've seen in Yellow House Canyon. By early August I've recorded less than 5 1/2 inches of rain on my place. There were no beebalms in June, and where hundreds of basket flowers came up last summer, one bloomed this year. I've seen almost no midsummer copper mallows or caliche globe mallows; even the afternoon-blooming sand lillies are hard to find this year. On the other hand, in a low spot where I burned last fall, devil's claws and giant Maximilian sunflowers, two native herbs whose seeds must have lain dormant for years, are growing where they've never grown before.

In wetter years, the high drama of summer on the Llanos takes place in the afternoon skies from May to mid-July. For weeks on end the prevailing winds blow moist air up from the southeast. All along the front of the Caprock the canyons act as moisture tunnels, pipelines that funnel this warm, moist air into beachheads of dry air meteorologists call "drylines" atop the plain, spawning massive weather cells. Most are ordinary thunderstorms, but if the cell begins to rotate and sucks moisture up the canyon wind tunnels at ever faster rates, at some level of critical mass the entire cell will go spinning off to the northeast, trailing tornado funnels behind it. Canyons, escarpment, and prevailing winds make the Llano Estacado edge the great tornado-spawning ground of the Southern Plains.

Such storms roar in on you in this open country with slight warning. Across the canyon the sky grows dark. Minutes after, a yellow pall of dust hazes out details in the far wall. When the wind hits, it drives rain and hail horizontally before it; lightning bounds along the mesa tops. The din against the tin roof of my cabin is terrific, a mounting crescendo that peaks, rapidly falls off, and then often returns from a new direction as the Coriolis force causes the cell to spin and pound in again.

If it is still daylight, a summer thunderstorm on the plains can have a couple more acts. With the smell of ozone and soaked juniper heavy in the canyon, I walk outside to the sound of waterfalls—more than a dozen are active along the rimrock behind the house after a hard rain—to watch the sun drop beneath the cloud cover and flood the fresh-washed slopes in a yellow light that can almost be taken in the hand. The effect lasts only a few minutes, but while it does it is so exquisitely beautiful it hurts, so unabashedly lovely it is almost corny.

Next day the valley in front of the house is under water for hundreds of feet across, and for two weeks Yellow House is dazzlingly *verde.* Every visitor remarks on how green it is . . . and on the sudden proliferation of life, insect and arthropod and amphibian, that hatches fast and cycles through in almost desperate haste to eat and live and breed while the moisture lasts.

Even without a calendar you know when summer is fading by the return of red-tailed hawks hanging motionless above the south rimrock, riding the updraft from north winds; by the presence of inexperienced nestlings like the little Bewick's wrens that sit on the railings of my deck, or the young lark sparrows grubbing around like little quail almost underfoot; and by a general yellowing of the entire country as the summer grasses set seed and dry out. By the time of the autumnal equinox, as I run in the early mornings, I see snake trails through the sand on the road and, in the quiet evenings of this time of year, hear many little squeaks in the grass. The snakes are mopping up before starting their annual fall return to the south-facing rim and their winter hibernacula.

The slow creep of autumn on the plains is not as easy to appreciate as the color bursts of woodland in the fall; it's the difference between Andrew Wyeth and Jackson Pollock. In the one are nuance and shifts in value and in crispness of outline that are hard to see unless you train yourself to look; the other is like being hit in the face with a bucket of water.

The gestalt of autumn on the plains turns on all the switches in me, though. I stir with it in September, when the snakeweeds flower, laying strong yellows in the hollows of the slopes. Yellow is the operative color of autumn on the plains, something that dawns on you come early October and the first two hours after sunrise are electric with a yellow light that by November suffuses most of the day and seems to have been absorbed by the whole countryside. The first frosts reveal the extent of it. The soapberry trees go golden first; through October and early November the yellow spreads to the willows, to the grapevines stringing the hackberries on north-facing slopes, then to the hackberries themselves. The crowning yellow of the Southern Plains is reserved to the cottonwoods, and if it is a relatively still autumn and the first freeze is a hard, quick one, for a couple of weeks canyonlands cottonwood galleries can rival the aspen display of the mountains. More often, fall winds and light freezes mar the cottonwood performance.

Up north, in the Red River canyons, the sumacs, shin oaks, and mountain mahogany paint that country into a quilt work of reds, purples, and oranges in autumn. But in the prairie canyons there is not much hot color in the fall outside the ribbon of tamarisk magenta along the rivers. At low light angles, gossamer-white old man's beard seems to drape most of the sage and fences. By October 10 cranes flute and croak overhead and insects stage their last, frenetic hatches. Flies and wasps are everywhere, and their numbers make you aware that there are no grasshoppers anymore, the cool nights having run them down like little clocks, and that all that remain now of the hugely gravid mantises of late summer are scores of hard, ribbed little egg cases glued to tree limbs and fence posts. I mark their locations. In early spring I'll sprinkle a few around my garden plot and root for the home team when the little mantises hatch and start to prowl.

The colors of autumn are gone a month before the winter solstice, when the elms and tamarisks finally drop their leaves, but the temperatures and feel of autumn linger well into December on the plains. Winter does not descend like a curtain fall; it comes in bursts that draw closer and closer together. We call those bursts northers.

Like the living entities the Comanches believed them to be, plains northers swoop down with the rush of a predator. Atop the Llanos one can see them approaching for miles across the flat, but they cover ground with startling speed. In 1986 one hit Yellow House at a clocked 84 miles per hour, blowing tin roofs off farm buildings for miles around and dumping my tipi, the only time it has ever gone over. Southern Plains literature from Cabeza de Vaca and Coronado on is full of the surprised reactions of pilgrims hit by their first northers, including the disbelief of frozen cavalry officers when a Comanche band escaped a

A norther, churning up dust on its leading edge, swooping down on Yellow House Canyon

skirmish in Blanco Canyon by turning head-on into a norther blowing blue ice and snow at tracer-fire speeds.

But when a norther has blown itself out and dumped its snow load and the night sky clears to a frosty glitter, calm, bright winter days of incredible, Colorado-like beauty often follow, the sky bluer than at any other season, the canyons full of shadows because of the low angle of winter sunlight. I have always reacted strongest to what writer John Nichols has called "the last beautiful days of autumn," but the clear, sunny winters of the plains are hard to resist. I think, in fact, that the most beautiful I have ever seen Yellow House Canyon is under a foot of snow, the temperature a quick 10°F, stars like frozen teardrops hanging by the billions etching Turkey Mesa in blue-silver silhouette.

LIKE THOREAU, though, I sometimes feel regret and wonder at what is missing from my seasonal observations. The grasses and forbs I watch year by year seem oddly purposeless. Life does not exist for itself alone. For 10,000 years and longer these have flowered and set seed and competed with one another for light and space to provide for the greatest biomass of a single mammal species in North American history. Now the grasses on my little piece of the plains grow rank and heavy, annually preparing themselves for an event that no longer takes place.

It's the September buffalo migration into the canyon that is missing. With their water and protective rimrocks and waving expanses of

Canyonlands snowscape

grasses, Blanco and Yellow House canyons were the prime winter buffalo pastures on the Southern Plains, and the imagery of 10,000 buffalo arrivals is not an easy thing to forget, not for plants, not for people. I have one good mental image of it that plays in my mind because I've read a journal kept by Elliott Roosevelt, Teddy Roosevelt's younger brother, who experienced Blanco Canyon when its fauna were yet intact (specifically, the winter of 1876–77).

Little bluestem, Yellow House winter

According to Teddy, who wrote about the adventure in a magazine article for *St. Nicholas,* with six illustrations by Frederic Remington, in 1889, Elliott and a cousin, John Roosevelt, had joined a group of buffalo-running toughs who were bound for the heart of the Texas herds. They ended up on the upper Brazos in a place called Blanco, a "cañon-like valley" whose running stream attracted unbelievable swarms of deer, antelope, wild turkeys, and bison during the dry Texas winters. Actually, except for his wildlife descriptions, not much about Elliott's journal is compelling; it was the sort of western adventure thousands of well-bred young easterners and Europeans had in the West and often wrote about. But I envy Elliott Roosevelt. He heard the full symphony in a place where I have had to be content with strings but no horns, percussion but no woodwinds.

One morning that Blanco Canyon winter, camped beside a deep pool in the little river and keeping the fire while his companions hunted, Elliott heard the scuffling of heavy hooves on the cliffs above him and watched motionless as a herd of bison descended (for one of the last times?) into the canyon for water, along a trail so sunken that in places the huge animals disappeared entirely from view. One by one they waded into the pool, and as the cow in the lead raised her head to stare at the human camp, nostrils working and water dripping noisily from her shaggy muzzle, the sun broke over the rimrock and lit her steaming, chestnut winter coat. Then with a little flip of her tail she whirled and led the herd back up the canyon wall, their broad hooves making a sucking sound in the mud. In less than a minute they were gone, leaving nothing but tracks and pieces of hair hanging snagged on hackberry limbs and a smell that lingered on the still air for an hour.

I record for posterity's judgment that the individual who almost singlehandedly robbed the rest of us of such primal delights in Yellow House was named George Causey. Causey and his brother made permanent camp in Yellow House from about 1877 to 1882, used to climb the mesas here with a spyglass to scout the movements of the shrinking herds, claimed to have killed 40,000 bison. One admiring contemporary bragged that Causey had slaughtered more buffalo in this canyon in one winter than Buffalo Bill Cody had shot in a lifetime on the Northern Plains. No one much pays attention to George Causey anymore, but when they do he assumes Homeric proportions as a West Texas pioneer. Like J. Wright Mooar, who performed a similar function (and remained proud of it to the end) in the Double Mountain Fork-Muchaque Valley country, Causey, his society, and evidently its descendants believed he was killing buffalo to advance the cause of Christ.

The animals went first. Then their smell, and next their tracks. Their droppings melted into the soil when the rains beat down and the lingering tufts of hair eventually were blown away or hauled off by nesting birds. Their wallows, which in winter had given the country a cratered, moonlike appearance under slanting light and in summer had brimmed with wetland vegetation, gradually filled with organic matter and drifting sand until they disappeared. Their trails into and through the canyons deepened into gullies, detaching mesas from walls, finally becoming unrecognizable. Today the only physical evidence that the great animals were ever here is the occasional hackberry, shorn of limbs up to rubbing height, standing lonely on a slope. That and the infrequent skull eroding out of a stream bank are all that remain to testify that only a century ago this was the heart of the buffalo country.

These and the expectant, waiting grasses.

EVEN LESS that is detectable to human senses remains of the rich assemblage of animals that coevolved with the bison herds.

On a late winter afternoon in 1987, my dogs dug out a wolf den under an immense boulder in the box canyon back of the house. Even to them the scent was very old, very faint, maybe dating to the last century. My female husky, Kooa, excavated the opening, disappeared inside, and refused to leave for nearly an hour, such was her fascination. She has been around coyotes for most of her life, but her reaction to the smells lingering in the stale air and the dried feces of that long-buried den was striking. She was spooky, cautious, but reluctant to leave and, in an anthropomorphic way, almost sorrowful.

One of the extinct species of wolves from the Southern Plains had reached across three-quarters of a century. The wolf taxonomists Edward Goldman and Stanley Young believed that the upper Brazos was the approximate division between the two subspecies of wolves native to the Texas plains—the Texas gray wolf (*Canis lupus monstrabilis*) and the larger, lighter-colored plains lobo, or buffalo wolf (*C. l. nubilus*). Just as the modern bison descended from the larger Pleistocene species, these wolves descended from the giant dire wolf, which became extinct about 8,000 years ago. The wolves of the Texas plains were a vital part of the old plains ecological equation. Along with the native peoples, they had served to cull the bison population and act as checks on its high reproductive rate. They were intelligent, in a way that the brightest dogs never attain, and when their natural prey was exterminated they unhesitatingly turned to cattle and horses (colts were a special delicacy). Plains wolves outlasted bison by about half a century, a bit longer in the Canadian Gorge country than on the Texas side of the Llanos. But by the 1930s they were extirpated. Buffalo wolf genes may survive in a few captive animals; the Texas gray wolf is believed to be extinct.

Why? Why hunt such an admirable animal, one so closely intertwined to human mythology, human history, human evolution, until it is *erased?* The answers are old; I have no new insights. The artificial, transplanted European stock-raising system would not brook losses to predators, and the practitioners of that system controlled the levers of power on the plains. The charges that wolves were cruel and wasteful and dangerous were obvious rationalizations, although there was some merit in the rabies fear. Hydrophobia horror stories involving wolves were stock stories on the early plains, and plenty of them weren't made-up stories.

But, mostly, wolves represented the wild. Not only the very essence of the enemy that Euroamericans had warred upon in its every form since the 1600s—and wolves perhaps came closer to symbolizing the resistance to European occupancy than any other animal—but also the wild animal within our own natures. Hermann Hesse captured the idea as well as anyone in his novel *Steppenwolf:* "He calls himself part wolf, part man. . . . With the 'man' he packs in everything spiritual and sublimated or even cultivated to be found in himself, and with the wolf all that is instinctive, savage and chaotic." Killing wolves, thus, was an act of self-flagellation and release from guilt for Christians. In the late nineteenth century every hardware store had a shelf of strychnine and no one passed a carcass on the plains without lacing it with poison. It was almost an act of atonement.

Their dens were here, in the canyons. Rollie Burns, who in 1888 became the manager of Western Land and Livestock Company's holdings centering on Yellow House, found the side canyons in particular full of wolf dens. In 1892 he hired two full-time wolf hunters, who killed twenty-five here that winter. The wolves that occupied the den my dogs dug out could have been among them.

Naturally, Texas contributed heavily to the wolf war. The famous No. 4 1/2 Newhouse steel leg trap was developed in Texas. But it was on the Llanos that "wolfing" reached its most artistic expression. From 1873 to about 1885, bison hunters were rivaled in numbers by professional wolfers, who lived in dugouts in the Llanos canyons and made two to three thousand dollars a year poisoning wolves for their pelts. In the process they unleashed a chemical war on nature from which the fauna of the Southern Plains have yet to recover. Hundreds of thousands of animals, not just wolves but also coyotes, skunks, raccoons, bobcats, and cougars, were poisoned directly. Kit foxes became extinct on the Llanos. Eagle populations were decimated. The big plains ravens, once the most common scavengers of the region, lay silent around every carcass like black blood spewed from the belly of some great, gutshot beast. Only a pitiful few remain today. With strychnine, a Comanche chief in 1883 wiped out all the wolves in the valley of the Salt Fork of the Red in one week, skinning 150 of them for the market. No one knows how many mustangs, pronghorn, and bison were killed indirectly, for an animal dying of strychnine poisoning drooled and vomited, and any grazer that later ate grass coated with crystallized strychnine would die. Because they lost so many ponies this way, Indians hated wolfers more than they did buffalo hunters.

Perhaps the pioneers didn't believe they could totally eliminate the wolves. In Lubbock County, historical markers relate that 1870s pioneers saw thousands of wolves, marching twenty abreast the story goes, ascending Yellow House Canyon and heading in the general direction of New Mexico, an apparent self-exile. But the sons and daughters of the pioneers knew extinction was in their grasp and never hesitated. I've listened to a taped oral history interview in which a prominent plainsman and his wife described the death, in 1917, of the last wolf in the Panhandle. I grew up hunting, as an adolescent shot animals unthinkingly too many times to feel comfortable remembering them. Maybe the pioneers represented our adolescence as a society, because they recognized that "it was the last of that species, an extinct species" yet found the memory of it mostly funny.

IN *Wilderness and the American Mind,* Roderick Nash has the modern bioregional movement doomed to fall short of the ideal because life in natural settings today doesn't have bison and wolves and bears in it. He's right. It's not ideal. It should be wilder. Meantime, I'm a victim of the era that created this imperfect view of the plains where most of the big animals are gone.

The midsized animals—deer, bobcats, coyotes, turkeys—have come back from near oblivion, though, and the smaller communities are mostly intact despite much lower populations than originally. Prairie dogs and their predators, the extirpated little black-footed ferrets excepted, are mostly in good shape. There are plenty of big plains woodrats in the rimrocks, deer mice in the prairies. The tiny predatory howling mice, which actually do howl, sitting upright with muzzles pointed skyward like little coyotes, stalk grasshoppers and other mice in fine species health. What all this translates into is a lot of rodent biomass, hence a lot of rodent predation. A lot of snakes, in other words.

Snakes give most people the willies, never mind the macho West Texans and mystic Hopis who do intimate things with rattlers in the name of tourism and religion. The reaction may be genetic: mammal aversion to reptiles must go back 70 million years. Only a few of the reptile orders have survived into our own time, and most are the specialized ones that prey on mammals. One of the most specialized and highly evolved of all these is the rattlesnake, in this particular region, *Crotalus atrox,* the Western diamondback or "coontail."

Although such ecological connections seem not to have dawned on the Sweetwater Jaycees, who hold an annual rattlesnake roundup, coontails are mousers *extraordinaire,* responsible in large part for keeping rodents from overrunning the plains. But in addition they have a tendency to strike at and bite the odd horse, dog, or human. The truth is, despite high rattler populations in the canyonlands, not very many people get bitten. Caprock Canyons State Park has had only two snakebites in seven years and with more than half a million visitors. But the stories about those unwary or unlucky enough to get bitten are spooky. Billy Harrison, curator of archaeology at the Panhandle-Plains Museum, has told me his story. It was early spring, April, a warm day on a dig in the Canadian Breaks. Billy had seen a fresh snakeskin and was thinking that he had better watch himself when he felt a blow on the back of the calf, above the boot. He was in the hospital a month, laid up at home for six. Under the full load of poison, he hallucinated that the hospital attendants were in a plot to kill him. Where the peptides in the poison began to digest his flesh is a golfball-size cavity. He is scared shitless of snakes.

In any case, I live with them, as do lots of westerners. Four of my dogs have been bitten; one died of multiple bites. I won't kill a rattler in the wild, but, pragmatically, I shoot the ones that want to move in (didn't Abbey, despite the infamous "I'd rather shoot a man than a snake," finally kill the rattler under his trailer in *Desert Solitaire*?). That this approach has decreased the snake population is not borne out by the mathematics of replacement: for four summers I had to shoot between seven and ten rattlers a season right around, and in two instances in, the house.

What I have learned is to keep the construction debris cleared away, to keep the screens repaired, to watch what the hell I'm doing, . . . and to get cats.

It's the ecological approach. Yellow House snakes of all species overwinter on the south-facing, sunny slopes across the valley. When they disperse to their summer hunting grounds they use sensory glands on their tongues to test the air for rodent signs. An old country saw is that "snakes hate the smell of cats." Actually, cats wreak sheer havoc among the local mice and rats. If the experience of my fifth summer here can be trusted, it works. I've not seen a single rattler around the house. Of course I do periodically have to chase coyotes out of the yard.

It's the rule of unanticipated ecological consequences. Coyotes dearly love to catch cats.

FIRE is a natural phenomenon that plains people have had to learn to love. Prairie wildfires, scourge to the early ranchers, a part of the plains wilderness that was resisted and feared, we've had to recognize as a positive good, cleansing and regenerative and absolutely necessary for ecological health on the Southern Plains. Without which we'll end up inundated in a green sea of mesquite thicket and tumbleweed.

Because these, of course, were what natural and Indian-set fires controlled on the Great Plains. They kept this region, whose dominant vegetation under stable, or true climax, conditions seems actually to be low brush, a grassland empire. And more than anything else, even more than overgrazing, although that tragic blunder was also catalytic in the ecological change of the past century, the zeal with which fires were repressed allowed the native grasslands to degenerate to their present mesquite- and broomweed-infested condition.

The spring of 1988 proved that fires can still rage in West Texas, but if the old descriptions are reliable, modern plains fires pale beside those that roared across this country a century ago, before highways and cultivation created firebreaks. Lightning caused many of them. Indians were responsible for even more, burning mostly as part of war strategies but also to keep the plains grassy for game and their ponies, to burn up the snakes, perhaps occasionally to manipulate bison herds to graze near favorite campsites. Even cowboys set a few. In 1877 a cowboy trying to drive hogs out of a shinnery thicket in Blanco Canyon set a fire just as a hard southwest wind came up. Usually the canyons acted as the natural firebreaks of the Llanos country, but this fire engulfed Blanco Canyon before roaring off to the northeast. Four days later it was finally stopped by the sandy floodplain of the Red River. Hank Smith, the first Euroamerican settler in Blanco Canyon, credited it with having destroyed the best cottonwood and juniper groves in all the canyons in its path.

After three-quarters of a century trying to figure out why the Southern Plains doesn't look now the way it is described by the early explorers, and after chaining, bulldozing, and poisoning hundreds of thousands of acres trying to fight the spread of brush, West Texas ranchers are beginning to be won over to fire ecology by a scientist named Henry Wright. Wright and his colleagues specialize in high-tech burning. They use helicopters and helitorches and burn thousands of acres at a time on the big ranches east of the Caprock. And it's effective: burning stimulates the native grasses, fertilizes the soil, and done at the right time, just as the sap begins to rise in early spring, it will kill back exposed woody vegetation. To finish the job, range managers then recommend a heavy application of 2,4-D, Picloram, or Spike herbicides. Our chemical dependence is more pervasive than people ever dreamed.

In five years I set six fires on my place, most of them in March but a couple in the autumn to see what differences there might be in the way the land responds (fall burns seem to produce more herbs, spring ones more grass). The general conclusion I've come to is that, if you refuse to apply chemical poisons on your land, you have to keep burning it. All the natives—junipers, mesquite, hackberries, even plains cottonwoods—have evolved a strategy for coexisting with fire. They sprout back from the roots.

I have the greatest admiration for the lime green, oriental-delicate mesquite tree. Trimmed, a big mesquite is as much an iconic ornamental for a Llanos homestead as a cottonwood or a juniper. But the collapse of ecological health brought on by overgrazing and fire suppression on the plains has just created too many of them. Contrary to "cowboy ecology," which asserts it was introduced, the honey mesquite (*Prosopis glandulosa*) is a true native of the Southern Plains. (Just read Randolph Marcy's exploration reports from the 1840s and 1850s.) Its range has been expanding northward in response to a favorable climatic regime for centuries. Yellow House Canyon was at about the northern limit of its range 150 years ago; I've seen photographs of this canyon taken around 1910 that show virtually no mesquite here even then. Now the range of the mesquite extends past the Red River and up the Canadian as far as the Gorge. And it covers the floors of the Texas Llano canyons like a Hudson's Bay blanket. Established like this, the subsurface a web of mesquite roots that go down 50 feet or more, mesquite thickets crowd out most other species, aggressively drawing water away from grasses and other trees. A single fire will do little more than leave standing a mess of the best cooking wood in the world and start the thickets over again at ground level. Reburn the same ground about every three years (the natural burn cycle on the Southern Plains was three to five years) and eventually the spreading grasses will starve the mesquite root systems, or insects will kill the weakened plants. Or so the theories go.

When I bought my place, it was a head-high mesquite sea, overgrazed for decades. Grasses were almost nonexistent except on the upper slope. Russian thistle (tumbleweeds) grew along the fence lines, but it was about the only herb on my land. So I started burning. Every spring after equinox I gather a few friends and we burn fire lanes around the piece of ground that most needs a prairie fire. Then we start the burn on the upwind side. If the wind is over 10 miles per hour the conflagration quickly becomes an awesome roar with flames licking 20 feet high and yuccas popping like firecrackers. Within minutes a five- or six-acre plot will burn to what appears to be utter desolation. With the ground yet smoldering and the air still smoky, meadowlarks, roadrunners, and other birds flock to the new burn to harvest the exposed insects and lizards. Within a week their busy searches are taking place amidst green, inch-high shoots, sprouting from grass seeds that had lain dormant for years beneath the domination of mesquite and snakeweed. The millennia-old fire regeneration is a prairie miracle.

Fall burns, in particular, restore diversity to places like mine. Few of the modern natives know it, not even the Mexican immigrants, who use plants as medicines yet tend to buy packaged plants shipped from Mexico to local *curandero* stores, but a healthy piece of Llanos canyonland is a treasure house of medicinal plants. American pioneers were not in this country until late in the nineteenth century. An oral history project done locally indicated that, although these pioneers knew some medicinal plants, the species tended to be those of southern and border states. Hispanic respondents were more familiar with the southwestern species, but even that knowledge is not being passed down in our age of pharmacies and bottled, synthetic medicines.

My own piece of Yellow House Canyon is a tiny slice of land with nothing biologically remarkable to recommend it. But even in such a limited habitat it is possible to tick off a couple of dozen plants the Hispanics and Indians once knew and used extensively.

There are food plants in abundance. We have an impression of Comanches and Kiowas as pure meat eaters, and while their ethnobotanies did shrink when they moved onto the bison plains, actually they relied heavily on plants for food. The unpromising little hackberries, for example, were pounded into a paste that could be molded over a stick for roasting, a treat high in carbohydrates, calcium, and protein. Mesquite pods, 20 percent sugar (they taste like tart green apples to me), were stored in sackfuls by both the Comanches and the Apaches. Prickly pear fruits, cholla berries, and algerita berries were eaten raw by the Indians and made into delicious jams and wines by the Hispanics. Wild onions and mints grow in profusion along the draws; devil's claw fruits, cultivated by some of the tribes farther west, were eaten as wild okra. All of these are among the most common of the native plants here. They don't even dent the total plant food resources of the canyons.

In another category were the purely utilitarian species, like the Rocky Mountain junipers (there's one under the cliff above my house) whose trunks were cut and trimmed for lodge poles. Or the red-berried Pinchot junipers, used along with sand sage fronds as purifying incense in a variety of Indian ceremonies. Or the stripped branches of the native sumacs, woven into baskets and mats for 7,000 years by Indian women, the discarded bark smoked by the men. Or the yuccas, most famous for the luster-imparting shampoo made from their dried, pounded roots (hence "soapweed"), although the Indians knew that burning yucca seedpods make excellent flares. And the common grassland croton, an essential ingredient in a mix with animal brains for tanning and a tolerable insecticide.

It's the *materia medica,* though, the medicinal pharmocopoeia of the plains wilderness, that best demonstrates the empirical methods of Indian healers and Hispanic *curanderos* in learning their environment. With such knowledge, a 12-acre piece of canyon like this isn't just grassy slopes and cliffs that are a pleasure to look at but a medicine chest of cures. The list that follows represents a knowledge of nature based on hundreds of years of local intimacy with the plains and a very apparent empirical experimentation with the effects of various cures, several of which either remain in our modern pharmocopoeia or have set researchers on the path to discovering breakthrough drugs.

A look into the Yellow House Canyon medicine chest: (The remedies could be dangerous if you don't know what you're doing. Seek the advice of a *curandero/curandera* or an herbalist before trying them.)

BIRTH CONTROL. One of the most interesting of Indian medicines was the use of stoneseed or Apache tea (*Lithospermum incisum*) by Comanche women as an oral contraceptive. They swore that a daily intake of tea brewed from this plant inhibited pregnancies. Modern science, led to the development of the birth control pill by such Indian remedies, has chemically analyzed stoneseed and found hormone-like molecules that do interfere with conception.

CAST FOR BROKEN BONES. New Mexicans boiled the bark and branches of plains cottonwood to make a syrup that was poured over a broken limb to make a cast lasting about two months.

DANDRUFF, SKIN ABRASIONS. The Kiowas, Comanches, and Apaches all used the saponin, or soapy lather, in yucca roots and leaves to control dandruff and to soothe skin abrasions.

DECONGESTANTS, COLD RELIEF. New Mexicans used sage in boiling water. Comanches chewed the bark and swallowed the juice of sumac (*Rhus trilobata*). Most widely used, even by Anglo pioneers who picked it up from the Indians, were ephedra stems boiled to make a tea. The local species has enough of the drug ephedrine to be useful for decongestion plus have a stimulating effect like coffee.

DIABETES. The presence in mesquite pods of the natural gum, galactomannan, which slows the release of sugars into the bloodstream, acted as an inadvertent but effective treatment for diabetes for all those peoples who ate mesquite pods and seeds.

DIURETICS. The most commonly used diuretic on the plains was ephedra, which has a decided diuretic effect.

ECZEMA. Comanches boiled the flowers of the snake broomweed into a jelly and applied the mixture to eczema outbreaks.

EMETICS. To make a very effective emetic, Kiowas peeled and boiled the roots of the buffalo gourd. Comanches and Apaches chewed the fresh roots of the spurge, *Euphorbia marginata,* as both emetic and purgative. It is the Euphorbiaceae family which produces castor oil.

HEART AILMENTS. Sanapia, the Comanche Eagle doctor interviewed by anthropologist David Jones, used the sneezeweed *Helenium microcephalum* to regularize a palpitating heart. Crumbled under the nose the resin from sneezeweed flowers will induce violent sneezing.

HEMOSTATS, OR COMPRESSES, TO STOP BLEEDING. The Kiowas peeled the fleshy stems of prickly pear and applied them to stop bleeding.

INSECT BITES. The Kiowas crumbled the leaves of lemon beebalm (*Monarda citriodora*) into water for use on insect and spider bites; Comanches applied chewed sand sage leaves to the bites.

MOUTH ULCERS. Kiowas chewed the berries of various canyon junipers to relieve mouth ulcers.

RHEUMATISM. The New Mexicans boiled the leaves of sand sage and placed them in bathwater for rheumatism; they also applied boiled prickly pear pads to rheumatic joints.

SALVES FOR CUTS AND BURNS. The blossoms of star thistle (*Centaurea americana*), found widely on canyon slopes, were boiled by the Kiowas as a cut and burn salve. Comanches used latex from cut wild china trees to stem infection and promote the healing of wounds. The Kiowas also boiled Western ragweed into a wound decoction.

SEDATIVE. Silverleaf nightshade, a species of a genus known since Roman times as a sedative, was used by the Comanches as a sedative and general tonic. The roots were boiled and the decoction drunk.

SNAKEBITE. For snakebite, New Mexicans ground the root of the buffalo gourd and put it in milk to serve as an antidote to the poison. To draw out the poison, they made a poultice from spurge.

SNOW BLINDNESS, SORE EYES. Comanches used the ashes from burned willow stems to treat these ailments, through direct application.

SWELLINGS. The Comanches made infusions by steeping globe and copper mallows in water. The roots of gayfeathers (*Liatris punctata*) were chewed and the juice swallowed for hemorrhoids and swollen testes.

SYPHILIS. Jean Louis Berlandier reported in 1830 that the Comanches could cure syphilis in its early stages by using the gum oozing from mesquite on the lesions.

TOOTHACHE. Comanches chewed mesquite leaves. The Kiowas, however, knew something it took Western science longer to discover: they chewed the bark of the native willows to relieve toothache. And willow bark is where we first obtained acetylsalicylic acid—aspirin.

UPSET STOMACH. New Mexicans chewed and swallowed the leaves of sand sage for upset stomach. Comanches chewed mesquite leaves to relieve stomach acid and prepared a tea from the roots of wild buckwheat (*Eriogonum wrightii*) for stomach trouble. The Kiowas seem to have relied upon teas made from the wild mints for stomach problems.

YELLOW HOUSE CANYON is unfolding below me like a movie, a slur of green undulations. Two hundred yards back it had been invisible. Now it yawns open like "a well in the desert," as Albert Pike put it in 1832. Farmhouses nearly two miles away on the other rim had appeared to be only a few hundred yards distant across smooth fields. Now, running with the dogs on the south rimrock, I see my barn, framed by yellowing young cottonwoods, in the valley below. My colt—a yearling stallion—spots us instantly, stands transfixed. Already I can smell the crisp-autumn-flannel-shirt-jean-jacket scent of piñon burning in my wood stove. We sail, glide along the rim like hawks, the lungs and muscles having ceased protesting miles back, the body deep into that fluid, sustainable rhythm that sets the senses free to observe and the mind to organize and process. This must be how we once pursued game animals. There is an intriguing clarity about it. Coach and sports journalist propaganda aside, the real skills that sports teach and have always refined are predatory ones, and one of the reasons for our fascination with them come from our intuitive recognition that an athlete can do the things we have valued in our hunters and warriors for a million years. Football obscures least of all the prey-pursuer relationship, perhaps why it so rivets us.

We pull up short at a point thrusting out over the canyon, at a circle of rocks 20 feet in diameter enclosing several rock spokes. Eighteen miles distant, past billboards advertising Treflan and Roundup and Asana, blue-haired women in Buicks and Coup de Villes with "Shop Til You Drop" bumper stickers cruise Lubbock, looking for bargains. Down in the canyon I have mesquite coals going and a buffalo fillet thawing and a bottle of Llano Estacado red wine already opened and breathing.

IN FEBRUARY of 1877 a ragtag bunch of buffalo hunters from Rath City, over by the Double Mountains, surprised a group of off-reservation Comanches who were camped in Yellow House to hunt buffalo and, in what one of the former called "the most picturesque fight" he was ever in, managed to run them out. Until I set up a 16-foot tipi on my place in 1985, I think Nigger Horse's lodges were the last ones to go up in Yellow House Canyon. Later that same year I cut twelve willow poles, planted their bases in a circle in the draw just below the tipi, and bent the ends over and tied them with leather thongs to make a sweat lodge. A friend and I meantime had already built the medicine wheel—a rock season calendar marking solstices and equinoxes—atop the cliff overlooking my place.

Except that I enjoy being in and sleeping in and looking at a plains tipi, that I love the serene way I feel (the way everyone feels) after a sweat, that I experience a "something" I find rather inexpressible watching the sun rise along the spoke of the medicine wheel at summer solstice, I have no answer when visitors see these things and ask me . . . why? "Are you an Indian?" "No." "Do you want to be an Indian?" "I do not." "Then . . .?"

Then, . . . well, it is a connectedness. To what? To a simple, human technology that anyone can understand—and fix. To the intrinsically satisfying lines of the circle, the möbius. To the efficiency of the cone as an architecture for windswept places. To the smells of steam scented with sage and red-berried juniper and sweating human bodies. To the timelessness of the fine light of Jupiter and a naked brown body slipping into the cold water beside you to finish a sweat. To the feel of caliche pebbles under your backside. To the three hundred generations of human beings who experienced these same things in this same canyon.

Sensory only, you ask? No spirituality? Except for that one eternally cycling message of all the mystics, namely, that All is One, and that what seems separate and apart is merely a mechanical quirk of how we perceive time and space, I apprehend none. I'm not sure I need anything else. What was it Thoreau said? "We need pray for no higher heaven than the pure senses can furnish, a purely sensuous life."

Chapter 4

Los Cañones del Valle de las Lágrimas The Canyons of the Valley of Tears

ON A CLEAR and cold winter predawn I sit cross-legged in the dirt of Los Lingos Canyon. Alone in this remote place, my senses are whirring at the highest frequency I have been able to reach since doing a sweat with friends on the autumnal equinox two months earlier. The still, clean air is rich with scent. Below the falls, in the narrow sandstone gorge 30 yards from my camp, is a stand of willows, and under the heavy frost they suffuse this part of the canyon with the pungency of damp woodland, a smell that is almost a taste. Inside the circle of rocks that encloses my fire, juniper embers hiss steam and stored energy with every pop, this scent working a different register in the olfactory range, one that sings a harmony line to the earthy willow chords.

These and other, subtler, stimuli I take in with mouthfuls of biting air, and the gulps are occasionally so pure and value laden that, briefly, I wonder if I can taste the piercing light from Venus, too, and actually draw in two or three lungsful from that direction. Can you *taste* the dawn? This morning, at least, I have convinced myself that the dawn does have a taste and for once I succumb to what I fancy is a primal directive: affirmation of existence through sensual consumption of the world around me. For more than an hour, experiencing breaking day from a remote camp on the headwaters of Texas' Pease River, I try for the antipode of meditation; the goal is not transcendence of the earthly but immersion in it. Each physical measure—the smells, the cushion of loamy sand under me, the visual symphony of the evolving sky, warmth and frostiness, the rhythmic bounce of water as Lingos Falls waxes and wanes in time to some pulsation that retreats around the corners of my awareness—I strive to consume and to memorize for future retrieval. This is what Thoreau must have meant when, from the summit of Mount Katahdin, earth and rock beneath his feet and wind in his hair, he shrieked, "Contact! Contact!"

Q. What kind of place is Quitaqua?
A. It is one of the eastern brakes of the Staked Plain, a broken country, great many deep canyons. . . . It was a kind of headquarters for the Indians.

Comanchero José Tafoya, 1882

No way of thinking or doing, however ancient, can be trusted without proof.

Thoreau, *Walden*

OPPOSITE PAGE: *Portals into the sandstone marvel of Los Lingos Canyon*

Sunrise. A large raptor with spread wing tips and a wedge-shaped tail with a white stripe extending across it swings overhead, curious at the smoke. Curved yellow beak reflects reddish eastern light. It's a young golden eagle, common in this part of the Llanos canyonlands. The scattered juniper woodlands surrounding me in every direction soon sharpen into etched detail. Peeling, stringy bark glows with backlighting; clusters of berries, a dull magenta only moments before, glitter as scarlet as Christmas decorations as shafts of horizontal color laser through the canyon. There is the sound of dripping as the frost melts. A mist curls over the willows, maximizing their pungence. Gradually, absorbing this, I grow aware that my fingers are aching and realize that they are supporting most of my weight, that I am leaning forward tensely on my hands, straining as hard as I can to hear . . . what?

A DECADE ago, long before I had heard of Los Lingos Canyon, someone put into my hands a volume by a biologist named Paul Shepard titled *The Tender Carnivore and the Sacred Game.* I think I and perhaps a considerable segment of the environmentalist subculture were waiting for that book, or one like it. Building on the premises of anthropologists Claude Lévi-Strauss and Marshall Sahlins, whose various works challenged the Hobbesian view of early hunting peoples, Shepard eloquently evoked a neo-Rousseauian noble savage for the self-comparison of modern-day people. He argued that we humans had been on the high road of our evolutionary journey until led disastrously wrong by our abandonment of hunting and gathering, and a life in wild nature, for pastoralism and agriculture, which Shepard equated with the real fall from our original state of grace. Far from being uncivilized savages, according to Shepard, as hunters we led a rich and blessed existence, sensually and philosophically more satisfying than our lot today, reflecting as it does 20,000 years of estrangement from the wild.

The primitivist message had appeal in late-twentieth-century America, particularly among my generation. It was a time when historical injustices were common topics, and there was a long tradition in high and not so high literature, from Mary Austin to Gary Snyder to Carlos Castaneda's wondrous (if probably fictional) adventures with a Sonoran shaman, portraying Indians as counter-cultural heroes. Some affected an "Indianness" that would have been silly had it not been an almost pathetic grasp at authenticity and rootedness in nature, things the dominant culture seemed determined to eradicate. However urged by Indian activitists like Vine Deloria and John Fire Lame Deer or shamanistic teachers like Harley Swiftdeer, it was almost impossible even for ethnic Indians to function as an "Indian" in modern America because, in its authentic sense, Indianness means a pagan world view and an economic lifeway based on living directly from nature. Both these traits stand in opposition to and far outside the capitalist and Judeo-Christian paradigm with respect to human beings and nature.

What they knew of the Indian nature religions had influenced thinkers like Thoreau and John Muir and many of the early-twentieth-

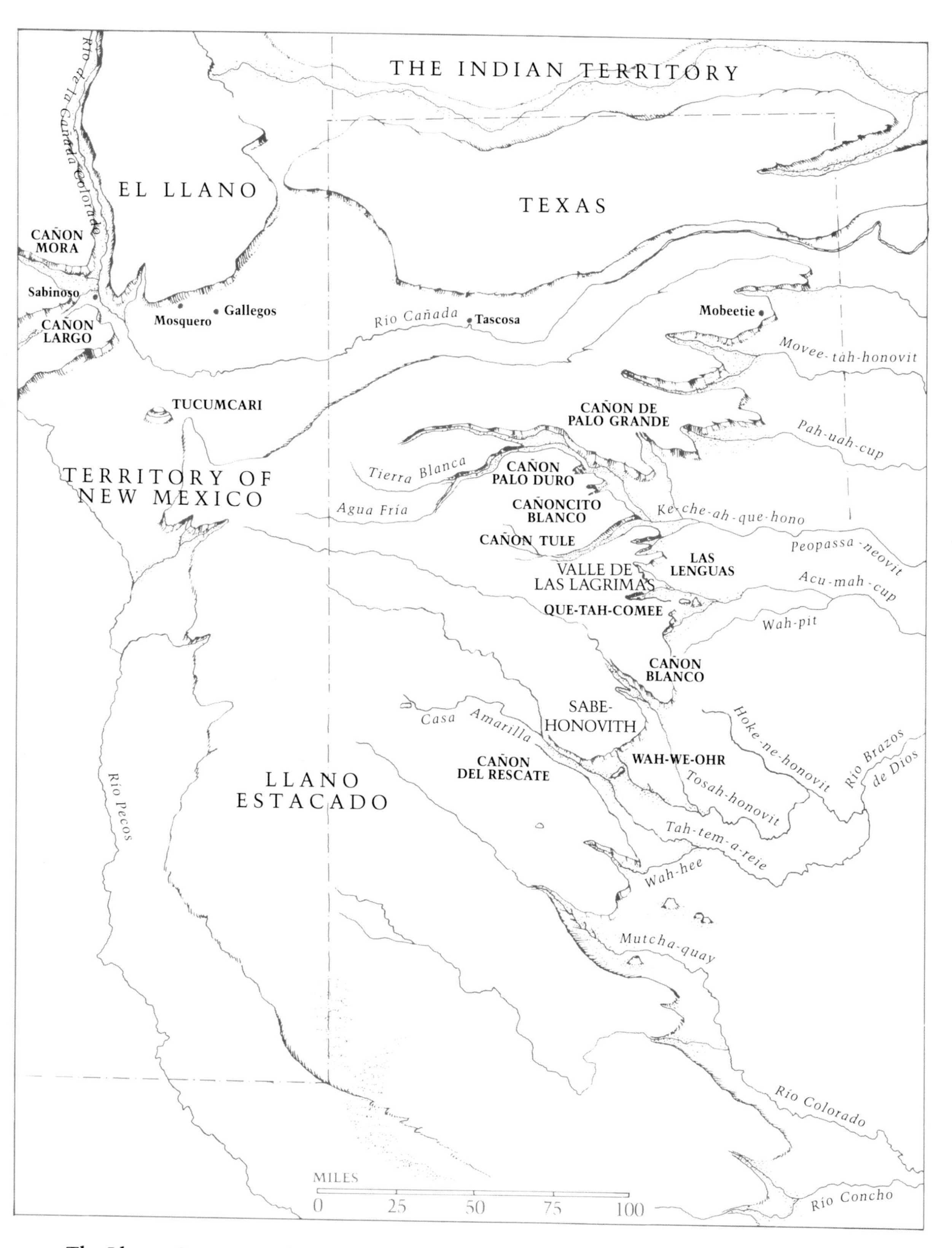

The Llanos Country in the Nineteenth Century with New Mexican and Comanche Placenames, FROM WALLACE, 1978

century conservationists, so it is not surprising that environmentalists would seize on Indians as ecological examples. There was little real study of the matter; it was just asserted that models existed, probably for every region of the country, for a sustainable, nonexploitive, harmonious coexistence between humans and the land. As environmental philosopher George Sessions put it, "primal societies throughout the world practiced a spiritual 'ecological' way of life. . . . This 'ecocentric' religious approach accounts for their cultural success for thousands of years and can provide modern humans with historical models for the human/Nature relationship."

On the Southern Plains our examples were, a few generations ago, uprooted and exiled to the last man and woman. Here, when people think of Indians at all, a more traditional image lingers in the form of the frontier morality play. Stories about Indians are fun now that they're gone, and old camps and burial sites are avidly trashed for frontier souvenirs, but in West Texas, particularly, Indians occupy about the same niche as Russians or grizzly bears. Ask a West Texan whose farm or ranch stands on the very ground where Comanches put up their lodges about Indians. I have. It is rare that the response isn't pure nineteenth century. Depending on whether he or she is an agribusinessperson first or Christian first, there will be an impassioned defense of one or the other, perhaps coupled with some sympathy or historical interest. But "inevitable progress" and "God's will" still play in the hinterland.

And yet as I write this the Comanchería has been snuffed out for only a century. The intensity and flair of the Indian occupation of this country are yet to me, sitting by a fire in Los Lingos Canyon, so recent as to resemble a great sound that has abruptly ceased at the very instant that we have turned to listen. Can 12,000 years of memories and myths and experiences with a landscape have been so completely scoured within a century? Is there truly nothing here worth knowing about, no lesson other than the ethnocentric one the present society draws from it? Does the landscape endure alone? And can we say why it now stands mute, when so recently it sang for other ears?

IN SOUTHWESTERN OKLAHOMA, outside towns like Cache and Fort Cobb and Gotebo, Kiowa and Comanche boys cruise the farm-to-market roads that follow old bison trails through rolling red and green plains extending out from the Wichita Mountains. They wear Levis and Wranglers and drive full-sized Ford pickups with bumper stickers advertising "Comanche Bingo!" Some of them—this you can tell if you watch their eyes for a lingering moment as you pause at stop signs in these small towns—are fighting boys, plains warriors still, who now fight among themselves for vaguer reasons and do not remember much the true cause.

On Saturday nights in this country there are still peyote ceremonies, combination Christian service and pagan ritual, originally introduced among the Southern Plains tribes as a cultural revitalization, similar to the Ghost and Sun dances on the Northern Plains, at a time when they were losing their homeland along the Llano Estacado. Peyotism and the

OPPOSITE PAGE:
Rocks, shadows, lichens, Los Lingos Canyon

Native American Church have lasted a lot longer than Ghost Dancing, though. While the Southern Plains people introduced the peyote cult to most of the other tribes of the West, today peyotism is far more widespread among the groups who still hold reservations than it is among those who don't here in Oklahoma, where Comanches, particularly, are often angrily divided over "Americanization."

Nonetheless, Saturday nights still take on an eerie quality when you drive the blacktops and dirt lanes north of Lawton. When moonlight glitters on shallows in the North Fork of the Red River and silhouettes the kopje-like piles of granite called the Quartz Mountains, I think of peyote and long-ago memories of the loose sand on the Mexican side of the Rio Grande and the green, pulpy, stomach-churning taste of cactus. And I fancy that I can tap into their visions, brain waves forming ancient geometric designs, their colors rich and earthy and ever mutating, which emanate from those Saturday night ceremonies and spread like pooling smoke along and over every contour of the plains.

Meantime, Los Lingos and Quitaque canyons and the Valle de las Lágrimas—the Valley of Tears—reflect the same moonlight in the state of Texas, clean and vacant and fenced and off limits. Three Kiowa women come to Bill Brown, then curator of ethnology at the Panhandle-Plains Museum in Canyon, and ask him to take them into one of the few canyons they can get into (Palo Duro), which he does, and they burn juniper incense and play a cassette of Kiowa songs. Bill grips the steering wheel of his old Nash and stares directly at the park road but is aware that one or two of them are mourning quietly, and here it is 1986. What he dreaded most, he said later, was showing them what could be interpreted as an exultant historical marker at the park road turnaround.

"LONG AGO, IT IS SAID." Thus began most of the Comanches' mythological stories.

"Long ago, it is said."

It was not long ago at all, of course, in rock time or in the kind of increments that span the birth and death of a genus. But for a people who marked out their history chiefly from oral tradition, the migration had taken place so long ago that many were unaware it had ever happened. It had, though, and its meaning was that the Comanches, no more than the Euroamericans who displaced them, were not "indigenous" to the Llanos canyonlands. They came late: 300 years ago, when the Pueblo Revolt set in motion the string of events that would remake the culture of western Indians and bring the Comanches to the Llanos, few if any Shoshonean-speaking peoples had seen the Southern Plains. For the next two centuries, band after band of hunter-gatherer Shoshoneans eschewed bipedalism for mounted quadrupedism, turned their ponies' noses south and emerged again in the Llanos canyonlands as Komantcia—Comanches. Like the American southerners who would take their country from them, they were expansionist invaders who drove less successfully adapted people before them. Along with the Kiowas, who were probably an old Puebloan trading group making a similar adaptation to the new conditions, these are the native peoples who provide the Southern Plains its most recent ecological example.

Who they were is a question too important to be ignored; not only who they were becoming as the country and their time affected them, but who they were originally. Who they've become to us a century later might be the best starting point.

Probably because of the protracted and chillingly efficient resistance they put up against their cultural genocide, the Comanches have been less romanticized than almost any other Native Americans. There are Indians in America—the Cherokees and Nez Percé, who suffered forced expulsions and subsequent tragedy; the Cheyennes, the "Beautiful People" of the plains, who have inspired so many romantic literary efforts; more especially, perhaps, the Hopis and other Puebloan peoples of the Southwest—whose philosophies have been held up by a variety of social critics as native models, whose experiences have been used to mirror the dominant culture's defects. Not so the Comanches.

Vast as it is, the Comanche library tends to proffer a consistent picture of a different sort, of an anarchistic mob of Hell's Angels on horseback who terrorized honest Christians, did unspeakable things to fair-skinned women, and in general contributed nothing to southwestern civilization except to serve as a particularly graphic reminder of the brutishness from which civilization has liberated us. (Indeed, a few modern Comanches relish the image: a group of European tourists taking in a museum in West Texas were told by the curator that, although the Sioux might be the more famous, "our Texas Comanches were really the worst and meanest Indians." There was a light tap on the curator's shoulder and the broad face of a Comanche man, who with two companions had been inspecting an exhibit around the corner, beamed at the group. "And don't forget—we still are," he said.)

Even the novelists, surely a more iconoclastic group than the symbol-oriented frontier chroniclers who wrote most of the histories, have been hard pressed to ennoble the lords of the Southern Plains. Neither James Michener's *Texas* nor Larry McMurtry's *Lonesome Dove* has much good to say about the Comanches except that they make good character builders for their Euroamerican protagonists. Max Crawford's *Lords of the Plain* is more Viet Nam era modern: famous Indian fighter Ranald Mackenzie is named McSwine in the novel and is an Anglo antihero rivaling Thomas Berger's Custer. But Crawford's Comanches are defective, tragically victimized caricatures. Lucia St. Clair Robson's *Ride the Wind* is a romantic novel, but most of the romance clings to her Cynthia Ann Parker character, whose captivity saga reads as if she were the blonde Ayla of the Comanche Cave Bear clan. The most sympathetic and culturally accurate modern fiction on the Comanches, Elmer Kelton's *The Wolf and the Buffalo,* while historically strong and gripping as a story, carefully sidesteps the question of philosophical legacy. So does Frank X. Tolbert in *The Staked Plain,* a book that is otherwise a gallant stab at realistic western history. In fact, Charles Webber's *Old Hicks,* written way back in the Romantic Age of the nineteenth century, is the only novel I've found that tries to make the Comanches into nature gurus. Literary critic Henry Nash Smith has one word for this book in his *Virgin Land:* "unlikely."

So what to answer when, as we are discussing Comanche religion one afternoon in the Quitaque Valley, a friend who wants to believe that Indians were nature's gurus but has read some of those Texas horror stories from the longest Indian war in American history says, "But did they really cut off arms and legs and throw live torsos, screaming, onto fires? How can we admire or learn from a people like that?"

It's a good question from someone who wants Hopis but has got Comanches. Not even high literature can help. Mark Twain, for instance, begins his essay on the French and the Comanches with this line: "Now as to cruelty, savagery, and the spirit of massacre."

Brrrp. Brrrrp. Brrrrr-rip. A bat, suspended by one foot from the roof of the cave just above my head, is making an odd, vaguely agitated sound that seems out of synch with the rise and fall of cicada song and the call of a dove in the canyon below me. I am sitting in the cool, loose, dirt of a cave hollowed from the sandstone walls of Los Lingos Canyon, escaping the August heat and idly watching a mule deer 100 feet below. It's a good vantage, as this vocal bat, at least one mouse, a pair of wrens, and dozens of incised names—names like Gloyna and Stidham, mostly dating back to the 1920s and 1930s—attest.

Like a lot of good hiding places, it also seems to inspire a certain cogitation. For instance: how long has this place been called Los Lingos, or some variation thereof? The original name must have been Las Lenguas, meaning in Spanish, The Tongues, because of the striking variety of languages spoken in this canyon by the traders who assembled here. Most of these canyons had acquired the names they now bear by 200 years ago, but this canyon was special and was probably one of the first of the plains canyons Spanish traders knew. In any case, names come and go. The Kiowas, less impressed by languages than by their successful attack on a hapless group from the party of Texans sent to "liberate" Santa Fe in 1841, thereafter called the stream American Horse River.

My friend Mike Long had pronounced the name more like "Linguish" this morning as we descended into the canyon by way of a 50-yard sandstone labyrinth, some 20 feet deep and in most places no wider than one's shoulders, locally known as the Linguish narrows. The narrows is a stunning place, too cool and lovely to have had to ourselves, and midway through our stay an active young diamondback let us know that it resented the intrusion. It was altogether too much like being in a closet with a snake to me, but Mike had other things on his mind, and with scarcely a pause he pinned the rattler's head with his walking stick, and we stepped over him and went on.

Out in the sunlight the morning air was yellow and still. We had come down a side gorge into the main canyon, which tracks a two-mile serpentine channel through red-gray sandstone at the bottom of the canyon V. There is a remarkable variety in the appearance of the Llanos canyons, but once again I was struck by the fact that, isolated from anything remotely resembling them by hundreds of miles, here was the quintessence of the Southwest: clean red cliffs, green junipers, the cobalt bowl of the sky, gurgling transparent water.

OPPOSITE PAGE: Descent into the labyrinth of Los Lingos narrows

Our destination was a somewhat famous waterfall, considered at one time or another for the National Register of Historic Places and as a

park in the Texas state park system. Because of landowner opposition, Lingos Falls today is not enrolled in either program. And for the same reason it's not easy to get into nowadays. For years the owners opened it to campers and swimmers, who responded by strewing it with litter, pulling guns on the foreman, carving graffiti onto every flat sandstone surface. After three Texas Tech University students died in the canyon in 1960, the posted signs went up in Lingos. That's where Mike had come in.

Mike and his brother, Jay, whom we haven't seen since the three of us left the truck this morning, are Californians who came to West Texas with their parents in 1960. Their dad, Beryl Long, discovered that he slept far better working a farm near Silverton than working on ICBMs in Los Angeles. Now Jay is a parasitologist in Amarillo and Mike, after getting a degree in range and wildlife management from Texas Tech, teaches science in Silverton and lives on their original 900-acre place, which he is restoring to its native grass cover under the new federal Conservation Reserve Program. I've known Mike for awhile and found myself peering a little incredulously at the redneck image he now cultivates. It's not easy being an environmentalist *and* a liberal Democrat in a West Texas town of 700 people, but after watching him interact with the local folk I had to admit he seemed to have it figured out.

Ten minutes downstream from the narrows was Lingos Falls.

By Rocky Mountain standards, by Big Bend standards, Lingos Falls is small-scale stuff. The waterfall itself drops perhaps 12 feet into the pool, which is not as large as a tennis court. But, as with so much of the scenery in these plains canyonlands, it just doesn't matter that there are more dramatic places somewhere else. Set a blue-green pool of water in a protected canyon cove in any desert country, add a waterfall, and sensual delight and surprise take over at once. What you do is, you strip and dive in. Which is what we did, and it's something we most likely couldn't have done at those other places because like as not there would have been too many other folks around.

Then we dressed and hiked down to the cave, picked through the flint scatter above it, reflected for awhile on how the valley downstream must have looked at night 125 or so years ago, with hundreds of Indian campfires dotting the flats like fireflies on a pond. Now, during a pause in the bat's chatter—which I'm pretty sure is directed at me—I hear footsteps on the roof of the cave and then Mike's voice.

"It's hotter'n hell. Let's go back down to the Falls."

Ten minutes later I am lolling on my back in the pool, admiring again the transparency of the waters. Between long pulls from the spring under the overhang Mike is talking—about mutual girlfriends we'd had at the university, about the way the Falls looks in winter, when columnar ice stalactites form beneath the overhang and glitter like crystals in the slanting sunlight, about how the spirit of the Indians seems to haunt this canyon when you camp here and watch the eagles and listen to what he calls "laughing birds" (canyon wrens). Because it occurs to me, I mention to him that this place had once been considered as a possible state park.

He finishes a long pull from the spring and looks at me as if I'm crazy. "Good God!"

An hour later we've insulted the rattler guarding the narrows once again and are out, and there is no Jay waiting at the truck. Mike assures me that such disappearing acts are Jay's trademark, but an hour passes and Jay's absence gets worrisome. I keep thinking about the three Tech students, one of whom had just completed a survival course, who came into Los Lingos in February of 1960 and got caught in a mild norther. Evidently they had completely panicked. One was found on the canyon floor dead of exposure. Another had missed the trail, tried to scale a cliff, and crushed his skull in a fall. The third, the one with survival training, had managed to get out of the canyon, spotted a house and was within a few hundred yards of it when he came to the rim of the side canyon we had descended this morning. About where our truck was now parked he sat down and gave up. For some reason he had failed to notice that the gorge, although deep and sheer, could easily be skirted.

We are out on a point, hallooing down into the canyon, when Jay staggers up to the truck, clutches at the water bottle, and collapses without a word against a tire. We look at him. He looks bad.

"You all right, Jay?" A nod, but otherwise nothing. "Fall and break something?" A shake of the head. "Turn an ankle, Jay?" Another shake. "Snakebite?" Not that either.

Finally Jay speaks.

"Lost. Lost all day long. Didn't realize we came in through a side canyon and got in the main gorge and couldn't figure it out. Afraid to drink the water in the creek. Guess I've overeducated myself on parasites. Hiked out to a windmill that was turning but it wasn't bringing up anything. Goddamned disheartening. Got in the creek to cool off, but couldn't remember where the springs were, or where we'd left the truck. Finally saw the house, way off to the south."

Mike and I look at one another guiltily, our bellies sloshing cold spring water.

Jay manages a weak smile. "Anyone interested in a large rootbeer in Silverton? I'm buying."

IN THE SPRING of 1986, as part of the state sesquicentennial hullabaloo and medicine show, the major university on the Southern Plains held a symposium to explain the "Llano Estacado Experience." Writer and critic A. C. Greene, historian Sandra Myres, and musician Butch Hancock were the celebrities of the affair. I had fled the state to the University of Wyoming in 1986, but I was asked to participate. Since an exhibit was to travel around Texas and was supposed to treat sequential cultures on the plains, someone was needed to make a few perfunctory comments about Plains Indians and the Llanos.

"Comanche Ecology" was what I decided to call my effort, and when I began pulling together the literature it got evident in a hurry that I was operating in empty space. Except for a single anthropologist, Jerold Levy, who a quarter-century earlier had outlined the problem in a brief piece called "The Ecology of the South Plains," very little research on Comanches and nature had been done. Writing from Washington state, Levy didn't appear to have much knowledge of the environmental setting of the Llanos, but he did sharpen the focus by pointing out that the

Lingos Falls in midsummer

critical factor in preindustrial human ecology on the plains was human population size matched against the carrying capacity of the land expressed as water and food resources. Thoughtful, although the conclusion—that droughts had limited the Llanos Indian population to a maximum of 10,500 souls—seemed to ignore a great deal of documentary information.

Sunlight . . . water . . . grass . . . bison. Wolves, obviously deeply intertwined in the mix. Human hunters in almost constant motion. Dogs, horses. Popular platitudes aside, how had it really worked?

I decided to start at the beginning, by taking the Comanches back to their roots as Shoshonean gatherer-hunters. I became familiar with some of the literature on human ecology, prey-predator relationships, bison evolution. Out of curiosity I got to know a smattering of general systems theory and its terminology, particularly the term "compatible system" (in human ecology a system or society adapted so harmoniously to its environment that lacking a major disturbance the whole is perpetual). An Indian specialist on the Wyoming campus in Laramie pointed me to the anthropological literature on human adaptations to new environments and new technologies. I read on Indian and, specifically, Comanche religion and philosophy. I discovered that the Freudians had done fascinating psychological profiles of Comanche "brigandage" culture. On my trips home to West Texas I did a lot of walking through the heart of the Comanchería, gazed at the old village sites, sought out the petroglyphs, and let the images sink in.

There was a song here not so very long ago, a new variation on an old song whose rhythm rippled across the countryside like the waves of time. Certain passages of it were familiar. But whence did it emanate? And how had it played before the crashing cymbal of that externally provided finale?

I presented my paper to an audience of about 200 well-dressed and well-heeled West Texans. They were polite, interested, maybe a little uneasy when I drew certain ecological parallels that reflected on modern Llanos culture. Then Butch Hancock got up and winked and grinned and took them into a half-hour cerebral maze on what he called "systems within systems within systems" (the organizers were wishing I had never made that remark about general systems theory) before picking up his guitar and strumming the familiar first chords of "West Texas Waltz."

READING, WALKING, THINKING, this is what I learned.

Long ago a great migration of Uto-Aztecan–speaking people swept down across western North America as far as the parrot-filled jungles of Central America. The gene pool of humans who spoke the dialects of this language family characteristically produced short, dark, bandy-legged folk with a tendency toward corpulence in middle age. At one time, the American West from Idaho to the highland valleys of Mexico may have been contiguously occupied by Uto-Aztecans. Eventually, other groups, Tanoans, Athabascans, surfaced like islands in a Uto-Aztecan sea. But the sea still existed in the West a century ago, and although Bannocks, Paiutes, Shoshones, Utes, and Comanches seemed diverse, and were, they possessed a commonality. No one now knows, because the names have not stuck, but thanks to a preserved journal from the 1870s, it is clear that the streams and canyons of the Comanchería bore names almost identical to the wonderful tongue-tiers that still cling to the topography and maps of Utah, Nevada, and northern Arizona.

Westward out of a mountain range in northern Colorado comes a river, the Yampa, and in great coiling canyons on the Colarado-Utah border it meets another river, the Green, which has snaked southward out of the Wind River Mountains of Wyoming. For hundreds of miles in every direction are cold sagebrush deserts and canyon badlands. Here for thousands of years, light years in space and time from condos and yuppie ski-slope chatter, the ancestors of the Comanches dug roots and surrounded antelope and communed with their pantheon of spirits.

The ancestral Comanches were not "plains" Indians, not bison hunters except occasionally. They were mostly desert canyon people who gathered plants, and because this economic activity could be done simultaneously with child care, it was usually done by women, giving them high status. The food search was almost unceasing and it took its toll, shaping a family-oriented society with an amalgam of kinship bands. In typical Shoshonean fashion, bands were named after the principal food source, the "eka" suffix so familiar in later Comanche band names meaning "eaters of." But no clan affiliations existed, even if (according to new Comanche scholarship) there may have been a vague tribal identification. Nor were there hunt police, warrior societies, or many ceremonies. Out of the remote past Comanches emerged as premier individualists, libertarians, almost anarchists.

Dimitri Shimkin, a close student of these ancestral Shoshoneans, found something else that interested me. Unlike many hunting and gathering peoples, these proto-Comanches lacked cultural mechanisms for limiting resource take. In other words, original Comanche society adapted to the carrying capacity of the land not by regulating animal or plant take but by regulating their own numbers. Birth-spacing through extended nursing of infants was one method. More common, more Draconian, was widespread infanticide.

A Comanche band of the 1870s would have immediately recognized their ancestral material culture. The typical wickiup survived in the form of the sweat lodge, and there seem always to have been tipis. Alone on the plains, the Comanches continued to use the ancient quadruped lodge frame rather than the tripod frame adopted by the tribes who approached the plains from the east. They also had dogs—large, wolflike huskies originally brought from the northlands—and a sort of travois so that with tipi and beast they could be desert, dog-powered nomads. They had little need for fast movement, because centuries of experimentation had given them the more than 150 medicinal and food plants that contributed to their phenomenal plant lore. Most astonishing, given later Comanche history, the ancestral groups were not only a passive people, they were literally patsies for more warlike neighbors.

They were religious, extraordinarily religious, and as the case with all of us if one digs far enough back, theirs was the oldest human religion.

The Comanche nature religion took an especially direct form, uncluttered by superfluous dogma. I continue to be amazed at what we once, all of us, instinctively knew about ourselves and our relationship to the earth, at least before Judeo-Christian theology told us that humans were the only species made in the image of God and before economic systems emerged that walled us off even further from nature by converting everything into exploitable resources. It has taken our very best minds to scale that wall, to peer beyond the shadows of our everyday reality and to really see. Yet the "pagan" Comanches never needed Charles Darwin to tell them that everything is connected in a great living web, Sigmund Freud to provide the insight that dreams lead us back to our animal selves, Albert Einstein to make them see that time and history do not proceed in a direct, linear march. From Asia and the dawn of consciousness, the ancestors of the Comanches had bequeathed a simple religious idea: there is a spirit animating the earth, of which the sun, "the primary cause of all living things," Comanches told early observers, was the self-evident origin. A feminine earth acted as the receptacle of energy from a masculine sun and converted that energy into growing, galloping, hissing, flapping, copulating life. The universal spirit pulsed through all the earth, was available in superannuated amounts in the form of personal *puha* (power, or medicine) to make things happen for those who could tap it. Men and women were relatives of, and sometimes assumed the form of, bears, eagles, bison. They partook of the specialized powers of the native (but never the introduced) fauna once a guardian spirit was acquired through an individual quest involving moderate suffering.

The goal of Comanche cosmogony was full immersion in nature, to see the human condition as yet one more expression of the impulses that motivated bison, coyote, eagle; to gain inner peace from the acceptance of full human participation in the wondrous cycle of birth, growth, death, a circle that one witnesses with each year's passing. Time did not stretch towards a coming, redemptive Kingdom of God on earth because for Comanches neither humanity nor earth had ever fallen. As was the case with so many Native Americans, only when Europeans appeared did the animals cease speaking and *puha* begin to fail.

History for the Comanches did not evolve progressively forward but rather cycled backward, via ritual and sacred places, to mythological time. Dance ritual was the principal technique for re-creating the mythological past; unlike the Kiowas and Southern Cheyennes, the Comanches are known to have participated in the plains Sun Dance ceremony only once. But there was preserved history in the form of folk tales, and like those of most Native Americans, these were not only instructional but also associated with particular landforms, so that both the stories and the very land itself were continually saying to people: *This is how one should act and live.* Perhaps most of all, Comanche oral traditions say, *see our kinship to all things!* or *see that this mesa, this canyon is sacred because of what took place here,* or because *especially here the universal spirit wells up and a sensitive person may easily become one with it.*

It was a beautiful philosophy, a poignant people's religion that was rarely under the surface. Observers noted that Comanches offered the first chunk of meat from a slain bison to the Great Sun, that they arranged bison skulls in semicircles around their villages to call the herds. Although no one left us a good description of it, in their dimly understood heyday the Comanches seem to have had some kind of bison ceremony at the heart of their religion, some sort of primal hunter's immersion in the thunderous power of their main beast. Comanche religion was, as D. H. Lawrence once commented about Indian religions, at its very best in "acknowledging the wonder."

LIKE BLUE NORTHERS, the once passive Comanches whirled through passes in the front range of the Colorado Rockies and swept onto the plains to their own hoof-pounding thunder. Horses to these hunter-gatherer Shoshoneans meant not just intensification of existing cultural patterns, but cultural transformation. Southward they came, first to be nearer the agent of their transformation, in the settlements of New Mexico and running wild on the Southern Plains, second to exploit the thronging animal wealth of that same country.

Wealth and power do things to human beings who have been poor. Caution and supplication get jeered aside. Confidence, swagger, a certain insensitivity replace them. The Comanches were no different. The Utes, who had first traded them horses and introduced them to Taos, all too soon found themselves whipped back into the mountains. Toward the New Mexicans and their Pueblo allies the Comanches grew increasingly contemptuous. But the group that experienced the full wrath of the *nouveau riche* horse hunters was the Plains Apaches, who had the double misfortune to have located themselves in the very canyons to which the Comanches, coming out of their Yampa–Green River country, were drawn and to have begun fixed agricultural settlements in them. Stream-side gardening villages in places like Blanco Canyon were soon Apache deathtraps.

Comanche success in seizing the Southern Plains and holding it against all comers for nearly two centuries was based on an elemental ecological fact. Acquisition of the horse meant that these formerly generalist hunter-gatherers could now specialize and dramatically simplify their economy. Ignoring their previous lifeway, the Comanches now hunted bison, ate bison flesh, drew bison petroglyphs, became one with bison. Never before had a Southern Plains people possessed the technology to specialize so, and the reason the Comanches prevailed is because they deliberately placed themselves on a lower trophic level than their rivals and exploited the available thermodynamic energy streaming from their sun god more directly and totally than anyone else ever had. It is little wonder that the most common Comanche name preserved in the old documents is some variant of *Isa:* wolf. Like the plains lobos, the Comanches were digging themselves into a narrow ecological groove, but so long as the herds lasted it was a potent groove indeed. And the herds were enormous.

Imagine turning one's back forever on jackrabbit drives and meager meals of roasted yamps and, horseback, gazing upon bison blanketing rolling grasslands, protein in unimagined and seemingly unlimited amount. No wonder the Comanche-Wolves became drunk at the prospect and exulted in their new condition.

It's not that this was a new occurrence on the Southern Plains. The modern bison—its reproductive characteristics, its migration habits, even its size—had been shaped by human predation from the animal's very origins 9,000 years before. The Llanos lifeway described by the initial Spanish explorations ninety centuries later, dog-powered hunter-gatherer nomads traversing the great sunlit wastes and exploiting bison for meat and leather, was so ancient that it had become built into bison population dynamics no less than the effects of wolves, weather, disease. The pressure undoubtedly intensified in the few hundred years before Comanche arrival, as the horticultural villagers located on the perimeters of the Llanos and effected a trade symbiosis, bison products for garden products, with the bison nomads. This hunting pressure in the core range, probably in combination with the great droughts of the thirteen and fourteenth centuries, explains why the bison herds had dispersed beyond the plains when Europeans arrived. But whatever the level of disturbance to the old human-predator–bison-prey relationship, it was nothing compared to the onslaught the horse Indians were about to make.

I'M NOT CERTAIN of the personality quirk behind this, but I seem predisposed to do stupid things on the plains in the dead of summer. Once I crossed 400 miles of Llanos to the Colorado Rockies on a clunker of a bicycle, packing 60 pounds of gear, in four days of 100-degree heat in mid-July. I knew better. I did it anyway. I also know better than to walk a railroad track in July, but here I am, shining rails curving beneath my feet, beckoning me into Quitaque Canyon. Somewhere I remember reading how Neal Cassady, the holy goof upon whom Jack Kerouac based the protagonist of *On the Road* and who in real life linked the Beat counterculture of the 1950s to the even more experimental one of the 1960s, died along remote rails in Mexico, counting crossties to win a bet. No doubt he never expected to die walking a track. I most certainly do not. Maybe I just like the smell of creosote in the morning.

Actually, I am looking for Comancheros—or the spirits of them, anyway. It is remarkably cool this July morning, still around 60°F at eight o'clock, and Tule is feeling good about it and about the way the three coyotes we jumped on our descent into the canyon showed him their tails. Up ahead the line of the Fort Worth and Denver railroad swings and disappears through sheer cliffs, and the image of the green valley through the portals of the track cut has me a little excited, too. This is my first time in Quitaque Canyon. I have deliberately come in midsummer because I want to understand firsthand why this canyon was so important in previous human history here. In essence I've come to experience Quitaque water, to see where it comes from, how it tastes. It is a small way to connect with the Comanchero Indian traders but an appropriate one, for it was water that brought them here, water and the certain knowledge that Indians with bison robes, sun-dried meat, stock, and human captives taken from ranches in Texas and Mexico were never absent from the Quitaque for long.

I also happen to be jumpy about walking a rail line just now. The lead-colored sky up on the plain behind us has started to heave and now begins the ominous, booming roll of thunder. As I am checking the storm's progress for the fourth or fifth time in as many minutes, off on the north rim about three miles away there is a jarring crackle, and quick as a rattler's strike blue electricity darts at the ground. And here my dog and I stand flat-footed, literally enclosed by steel.

Seconds later is another searing flash, brighter than a welder's arc, this time to the southwest and not two miles off, and we get the hell off the tracks.

Forty-five minutes later, the valley silent except for water dripping from hackberry leaves and the sound of a pour-off somewhere in the distance, we crawl from underneath our rimrock shelter. Tule, smart dog, spits out his share of the peanut butter cup I offer him and we resume our walk, track pumice crunching underfoot. Once, flying over this canyon, I had not been impressed, having looked down into Palo Duro, Tule, and Caprock canyons in the preceding hour. Now its attractiveness surprises me and I try to delight in it the way a troop of New Mexican Comancheros, weary from the monotony of two weeks of flat Llano, certainly must have.

They were part of a Hispanic expansion in the Southwest, these Comancheros, after two centuries of consolidation and acculturation in the valley of the Rio Grande, a folk movement of New Mexicans northward into Colorado and eastward onto the Llanos. The histories say that the Comanchero trade began in 1786 when the Comanches finally made peace in Santa Fe. The whole thing was much older, of course, the Comancheros (or *comanchieres*, as they actually called themselves) merely the last expression of the ancient plains trading caravans emanating from the Pueblo villages, and right down to the end many of the parties were Puebloan or led by Indians from the Rio Grande villages or from Pecos Pueblo. Along with the *pastores* and the *cibolero* buffalo hunters, the Comancheros made the logical assumption that the Llanos country was a part of New Mexico. For this they were regarded as heroes in Taos, Mora, Santa Fe, Nambe, and Gallegos, as trouble-making rabble in Austin, Houston, and traditional Texas history.

The canyons along the Caprock Escarpment were their favorite trading rendezvous on the plains, and it is Comanchero names, rarely those of the Indians, that later English-, German-, and Czech-speaking settlers preserved and happily butchered in pronunciation.

The names are about all they left.

The names and a few archaeological sites, like the one found in this canyon by a Plainview archaeologist in the 1970s, that prove the Comancheros were becoming permanent homesteaders in their favorite trading spots, settling the American West by trekking *east.* In classic New Mexican fashion they shaped the local Permian clay and native bluestem into adobe bricks, put up houses and cedar *jacales,* ran *acequias* from Quitaque Creek that were the first irrigation projects in the Texas Panhandle.

Texas cowboys, sons of poor frontier families who worked for men with big money and big schemes for more of it, young men who had more in common with the New Mexican homesteaders than they ever dreamed but saw only the differences, ran them out. It was simply one more expression of the race war that characterized late-nineteenth-century West Texas and that sent the Indians to Oklahoma and the New Mexicans whence they came.

Millennia-old habits are hard to break. As late as the 1890s New Mexican traders would not let go of the idea that with a few loaves of bread and items of hardware they could find Indians eager for trade in the Llanos canyons. Barbed-wire fences finally made them give it up.

Walking a track has its moments. Tule shows me a porcupine he has found and from prior painful experience knows to observe from a distance. We walk around an immense rattler, its head the size of a teacup, that probably attained its "immense bigness" (as an English colonist seeing his first rattler once put it) from never having encountered a human before. We come upon a snag nearly toppling from the weight of wild turkeys that crane incredulously at us and one by one flap off into the trees. We descend the canyon quickly via rail, but it's a little like viewing a country from a highway. Wasn't it walking a track (or was it riding a bicycle?) that struck Bertrand Russell with the conviction that he no longer loved his wife? Emerson said Thoreau hated the sound of gravel underfoot, yet there is good evidence that the Sage of Walden regularly followed the track going from his cabin into town.

I am able to tolerate five or six miles of the monotony of crossties and crunching gravel but no more, and when we begin to see the rolling plains and the Quitaque Peaks to the east, Tule and I bid farewell to the tracks and clamber down gray shale slopes to the creek. Except for an occasional pool, the upper parts of Quitaque Creek had been dry. But although narrow and relatively shallow, Quitaque Canyon evidently cuts into a dome in the Ogallala Aquifer and, from its numerous periphery gorges, springs trail water and corridors of verdant vegetation down to the main stream. Midsummer or not, the Quitaque is running, murmuring sweetly through carved sandstone chutes and over gravel bars, its color a translucent blue-gray, the taste like tolerable well water although a bit hard.

Reacting subconsciously to the geology, we've come down at the point where the creek slices into the Triassic formations, and heading back upstream as I'd planned won't do. Downstream is the more intriguing; downstream are rapids and, faintly, less a sound than a suggestion, something else, a singing like a breeze in a cottonwood except that there is no wind today. The topographical maps do not show a falls on the Quitaque, but the geology is right, and the sound, pulsing a bit now as the dog and I pick our way around bends in the drainage, seems less and less imaginary. I can't think of just where, but I have looked at an old map, I remember, that showed a "Quitaque Falls."

Quitaque Peaks, an ancient landmark

Wild turkeys

My dog, ready for a descent

In ten minutes, I am sprawled in the cool sand of a cottonwood-enclosed grotto, the spray from a beautiful little 18-foot waterfall forming into droplets on my naked, sweaty skin.

Later, as I ascend the canyon along the base of its north wall and idly watch the kites carving out hexagons in the blue plains sky, it occurs to me that I reached neither the Comanchero site nor the famous Quitaque Tunnel, the only railroad tunnel still in use in Texas. Another time, maybe, I'll camp at the falls and investigate both. Today, though, I am much more interested in what the engineers left alone than in what they built. And the feeling of old campsites and the Comanchero presence at the falls was so tangible that even the canyon wrens sang with a New Mexican lilt, and I felt no need to see excavated ruins.

THE OLD TRADE of which the Comancheros were the final stage was intrinsically important to the southern Plains, not just because the Comancheros were the vanguard of New Mexican expansion eastward. The Comanchero trade was one piece in the old human ecology puzzle on the plains.

The Comancheros named the bowl-like valley that erosion has formed at the mouths of Los Lingos and Quitaque canyons El Valle de las Lágrimas because they knew that here, more than anywhere else, they could find Indian camps, Indians willing to trade or ransom the hundreds of Euroamerican captives, who were not so much pawns in an evil and criminal economy as surplus in the great Comanche effort at population increase that was taking place between about 1700 and 1850. And that has its own importance in the equation.

Understanding the Comanche life of mysticism and emotional release on an ecological level forces one to deal with numbers, much as I dislike them. The first task is to dispose of those that are, or were, impossible. There never could have been 100 or 60 or even 40 million bison on the Great Plains. The grasses would never have supported that many. The best way to determine bison carrying capacity for the Southern Plains is to extrapolate from U.S. census data for livestock, and the best census for that is the 1890 one, after the industry crashes of the 1880s had reduced cattle numbers to something ecologically realistic but before most of the land was broken by farmers. That census indicates that, in the 240,000-square-mile Comanchería, the carrying capacity was about 7 million cattle. Native bison are much more efficient users of the prairie grasses than cattle—about 18 percent more—so the prehorse Southern Plains carrying capacity for bison must have hovered around 8.2 million animals. Thus the entire Great Plains would have supported perhaps 24 to 25 million. The rapid spread of horses across the Southern Plains between 1690 and 1800 would have dented this number, though, because horses and bovines have an 80 percent dietary overlap, compete for water as well. Assume that there were 2.5 million horses (as J. Frank Dobie estimated) on the Llanos by 1800, and bison carrying capacity drops to about 7 million in the Comanchería.

That number is still incredible enough, but it doesn't convey the emerging equation, which involves coming to grips with the ecological factors affecting bison population, those affecting Indian population, and the cultural aspects of how the Comanches used bison.

The several, growing bison herds in the modern West provide some of the answers. For example, we know now that bison fertility is about 18 percent a year under normal conditions (a 51/49 female to male ratio with 70 percent of the cows breeding), and that the natural mortality rate without predation averages about 6 percent. In other words, fully stocked with 7 million bison busily grazing, copulating, birthing, the Southern Plains would have produced an annual increase of about 1.25 million bison. Endemic bovine diseases like tuberculosis, and very severe introduced ones like brucellosis and anthrax, in combination with fires, floods, and hard winters, were the principal nonpredatory enemies of bison. There were also strange and unaccountable die-ups like the one Charles Goodnight saw between the Concho and Brazos rivers in 1867, when a herd ate out its winter range, refused to move, and proceeded to fill an area 25 by 100 miles with carcasses. These kinds of natural controls would have eliminated roughly 420,000 of the increase. Leaving a surplus of 840,000 animals a year for predators. For cougars, who may have gotten a few calves. For grizzlies . . . maybe. Mostly for wolves, both canine and human.

For, of course, simple Malthusianism tells us that, unchecked, only two or three years of such increases would have destroyed the grasses. Bison evolution had been shaped around a certain high level of predation from the first. It intervened long before starvation to keep the herds in a healthy balance with the grass.

When Cabeza de Vaca, the first European to leave a written record, crossed the southern end of the Llanos in 1536,he commented that the trade in bison products between plains and Puebloan peoples was "vast." Other Spanish *entradas* of the sixteenth and seventeenth centuries estimated that the number of bison people on the Llanos compared favorably with the number of people then living in the pueblos along the upper Rio Grande, a population variously estimated during the seventeenth century at between 15,000 and 30,000. All of which is extremely "iffy," of course. But if there were as many as 30,000 bison hunters on the Llanos during pre-Comanche times, then we have something to go on: those folks obviously hadn't killed off the herds, for either trade or personal consumption. And if they pressured them, well then, mightn't human hunters have shaped into bison evolution over 9,000 years a remarkable ability to withstand predation?

Now came the Comanches, awed and worshipful of the bison, but with no tradition of organizational restraint on their use of nature and now sole owners of wealth in an abundance they had never known. On horses they came, in Technicolor and full splendor, prepared as no humans had ever been to take advantage of the unique ecological situation created by the Pleistocene extinctions—the enormous biomass of bison.

The Comanches were a conservative people. Until crushed beyond all hope by U.S. Indian policy, their ancient culture flowed with the inertia of a river. Like all of the three dozen or so tribes that became horse Indians on the Great Plains during that meteoric phase of the "Hyperindian" (as someone has called them), however, the move to the buffalo country changed the Comanches, in ways that were far-reaching in how they used nature.

One of the cultural changes that reverberated oddly involved a change in status for Comanche women. Psychologists who have studied old-time Comanche culture see an unusual amount of tension between the sexes. Put simply, like the characters in a Larry McMurtry novel, Comanche women seem to have been unfaithful in droves and with vengeance (at least subliminally) in mind. Buffalo-horse culture had overnight made the traditional plant-gathering activities of women less critical to survival than before; on the plains the Comanche ethnobotany shrank by 60 percent, to fewer than seventy plants. Hunting and raiding had meantime created a fine ego structure for Comanche men, one that more and more came to regard women as prizes and adornments. The women of many tribes recognized this process. There is evidence that some women actively resisted the movement onto the plains.

Women lost something else on the plains. They lost their old Shoshonean right to multiple husbands. Instead, Comanche men now took two, three, sometimes six or eight wives, a trait anthropologically associated with peoples whose populations are expanding. In the case of the Comanches and other buffalo tribes, there was an economic motivation, too. More wives meant more processed bison robes for trade. All the old Shoshonean mechanisms for controlling population growth seem to have been abandoned: infanticide also vanished along with polyandry. As a matter of fact, the Comanches stole women and adopted children taken in raids with an enthusiasm that leaves little doubt that the move to the bison plains became the catalyst to a consciously motivated Comanche population explosion. Although the estimates range as high as 30,000, six of the seven population estimates for the Comanches between 1786 and 1854 fall into a narrow range between about 19,000 and 22,000. Alliances with other tribes—the Wichitas in 1735, the Kiowas and Kiowa-Apaches in 1806, and the Southern Cheyennes and Arapahos in 1840—opened the Comanchería to a total population of, actually, around 30,000.

Bison wealth wasn't the only catalyst to a Comanche demographic transformation. Disease, introduced European diseases against which no Native American shield or animal helper or power amulet was successful, kept decimating them. Disease is the human disaster that complicates Comanche ecology, makes the final scenario an unfinished painting. Lacking it, Comanche numbers may have continued to soar, to 50,000, 75,000, or more. We don't know for sure because at least a dozen epidemics and pandemics swept western America after 1492. The 1816 smallpox epidemic is said to have killed 4,000 Texas Comanches; the 1849 cholera epidemic, perhaps as much as half their population. The anthropologist John C. Ewers has estimated, in fact, that disease cost the Comanches 75 percent of their population during the nineteenth century. AIDS may give us a small taste of what it must have been like to face measles, cholera, smallpox, influenza the way an Indian did. In Kiowa mythology, Saynday, the mythic hero, is told by "Smallpox" (who takes the form of a missionary): "No matter how

OPPOSITE PAGE:
Quitaque Falls

beautiful a woman is, once she has looked at me she becomes as ugly as death."

Fix on these critical statistics then. Between epidemics a bison-hunting population that built itself up to 30,000 or so, with an annual increase of 840,000 animals to share with other predators. What was the human share?

My friend Bill Brown and I have worked hard on this problem. Bill has had good success adapting a sophisticated model first worked out for caribou hunters in the Far North and based on caloric needs of a hunting population, plus the number of hides or robes needed for domestic use. Bill has estimated Comanche subsistence at about 6 bison per person per year, although the caloric requirement (in pounds of air- or sun-dried meat) was only about one animal per person. Bill's model not only gets us much closer to a historic plains ecological equation than ever before but also is borne out by at least one historical account. In 1821 trader Jacob Fowler camped for several weeks with seven hundred lodges of Southern Plains tribes in the northern Comanchería. Fowler was no ecologist; in fact, he could hardly spell. But he was a careful observer, and he wrote that the big camp was using 100 bison a day. At an average of eight people per lodge that works out to a yearly Indian individual consumption of around 6.5 animals.

Multiply the figures and Comanche-Llanos ecology begins to make sense. Take thirty thousand bison Indians. Factor in caloric and domestic requirements. Add a robe trade that seems to have averaged around 50,000 a year. Add 20,000 or so to account for the number of bison that New Mexican *ciboleros* and Pueblos took from the Llanos. Out of the seven million bison on the plain, the local Indian hunters were taking only about 260,000 animals a year. Equation: natural bison mortality = 6%; human predator share = 3.6%, leaving us with a figure for the long-sought wolf predation share at 8.4 percent or 590,000 of the little red-yellow calves a year. This is how the centuries-old Llanos balance worked, maintaining healthy bison, healthy grass, healthy wolves, healthy human hunters.

The Comanches might have been able to upset the balance. Like the Sioux on the Northern Plains, they were expanding, growing into their new country. The Comanches were a people who felt the sacredness of place and intuited through their senses a radiating kinship among creatures. However respectful they were of the bison, though, however remarkable their religion in "acknowledging the wonder," there is not much evidence that the Comanches were consciously limiting their use of nature. Indeed, they believed that bison could never be made to disappear by humans. Their war on other hide hunters was a war against trespassers, and they continued to insist to the end that the great bison herds had their origins underground, would emerge undiminished every spring from a great canyon rent in the surface of the Llano Estacado.

Lacking disease epidemics, would Comanche population have outgrown the equation? It's a question whose answer was never allowed to play out. A computer simulation projects that the pre-1850 Llanos could have sustained a bison-hunting Indian population approaching 60,000 (a little less than half Amarillo's present population) with no adverse effects on the herds, although competing predators would have suffered. Beyond 60,000 the Indian preference for breeding-age cows would have induced a population spiral in the herds and the history of the Paleo big game hunters may well have been repeated.

All of which is speculation, since it didn't happen that way. But fun and intriguing, and maybe the whole exercise is instructive as more than just a jigsaw puzzle if the future holds what some of us think it might. So here's one more guess. At sometime before that 60,000 threshold, particularly for horse Indians (for horses have ecological requirements of their own), water limitations may well have put a ceiling on Indian population growth. Hence, those who argue that water, not bison, was the limiting resource on the Llanos, take your bow. I'll bet Jay Long wouldn't disagree with you. Indian fighter Ranald Mackenzie wouldn't, either. Recognizing how important those canyons and their water were to nature's children, he wrote in 1872 that "there was a belief, a few years ago, that the edge of the Staked Plain would not support for any length of time a large body of Indians. This is a mistake as there is no country . . . better adapted to all the wants of the Indians." The larger the Indian population, the more horses they owned, the more critical Llanos oases like Los Cañones del Valle de las Lágrimas became.

SITTING beside a campfire in Los Lingos Canyon at dawn, what I strained so hard to hear, I have come to think, was a song not as it was played by a landscape, but as it was heard by a people. The Comanches, whatever our popular perception of them, were a people who thought poetically and mythically about this country, who made it theirs by designating sacred spots on the landscape, who intuited their link with the energy that surged from ground and grass. Their wild, exhilarating life, wherein sunlight and water, bison and grass, hunter and horse were fused into a sort of signal, primal poetry, became the very quintessence of what a sensuous life could be. Western science may grasp complex ecological issues more effectively, but will our rational explanations ever allow us to hear the Song of the Plains the way the Comanches did?

Sunlight and grass, bison and hunters, women and earth. Wolves, playing their ancient role, in an ancient equation. The images—the *puha* of the thing—reverberate across the years, and there are those of us who cannot let it go. I cannot, I think, because as I perceive them the plains hunters somehow recognized what we have such difficulty with: our search is not really for the meaning of the universe but, rather, for the euphoria of knowing that we are alive and a part of it.

All America lies at the end of the wilderness road, and our past is not a dead past, but still lives in us. Our forefathers had civilization inside themselves, the wild outside. We live in the civilization they created, but within us the wilderness still lingers. What they dreamed, we live, and what they lived, we dream.

—Thomas K. Whipple,
Study Out the Land

We need wilderness because we are wild animals.

—Edward Abbey,
The Journey Home

Face to face with Tule, a place not easy to put out of the mind

Chapter 5

Wilderness Cathedrals

EXPLORATION AND WILDERNESS are peculiarly American fascinations, and the two are bound together in an intellectual colloidal suspension in our culture. "Lewis and Clark" is an American code phrase for "wild," with all its associations—unspoiled, beautiful, natural, seen for the first time. Never mind the ethnocentric dimensions; Indians obviously had altered and been gazing on the continent for centuries, and what seemed "virgin" was actually a landscape maintained for hunting. But the effect of what Theodore Roosevelt liked to call "empty sunlit wastes" is the kernel of environmentalism in the American West. Europeans and easterners worry about toxic waste and acid rain, carcinogens and public health. In the West, despite nuclear plants like Pantex and nuclear waste sites like WIPP and unimaginable pesticide applications, wilderness is the Holy Grail.

And it has become about as hard to find, particularly if "empty" is a defining value. The paradox of wilderness protection is that as soon as you call a place a wilderness, the "emptiness" factor begins to spiral into oblivion. This is why the true explorer's sense, a kind of Adamic illusion, is almost impossible in tourist parks like Yellowstone or the Grand Canyon and not easy to have in officially designated wilderness (the Pecos Wilderness above Santa Fe is the best example in the Southwest) that are too near large human populations. Yet, if it is emptiness that one seeks, a thousand out-of-the-way places in the West, some of them public, most of them private in Texas, have it in abundance. If emptiness is ultimately the defining value of wilderness in modern life.

Of course it is not. Visual and biological richness, along with certain concepts about the wild that are distinctively American, are higher values than mere absence of people. But we still want the explorer's paradise. And why not? The entire West was such a place only a century ago.

I have not experienced the Adamic illusion many times in my life, but there have been a few. Yet the most deeply felt such experience was not in Wyoming or Montana or Utah. It was in West Texas, on private land. It came on the concluding morning of a four-day hike through the lower end of Palo Duro Canyon and up mysterious, unimaginable Tule Canyon.

It was late October 1985, and it was a walk, looking back now, designed unconsciously around intimacy. Intimacy with the regional landscape, for one, but also a subliminal attempt at renewing the bonds with the woman I'd been living with for four years, whose career was taking her away (an excuse that made it easier to sidestep the relationship difficulties we were mired in). I should have known better. The first two days, descending the sandy, gypy, salt-encrusted Prairie Dog Town Fork, our moods reflected the near lifelessness of that harsh, baking country. We fought angrily enough the second morning that for three hours we took different routes. When we met at a windmill I could read in the sullen silence a litany of my sins for which I had no defendable answers.

We camped that night under a big cottonwood in the mouth of Tule Canyon, the full moon so bright that the veins in the cottonwood leaves were visible in the shadows on the tree trunk. Katie's truck was 14 miles upcanyon, waiting by the highway in the dramatic Tule Canyon Basin, a kind of miniature West Texas Monument Valley and the only part of Tule either of us had ever seen. The hike up the canyon looked doable in a day, although the map showed the last section as a series of deep, winding loops that began with a narrow defile the U.S. Geological Survey called the Narrows.

We didn't make it in a day. We barely got out by sunset the second afternoon. We didn't argue anymore, either.

I was, admittedly, totally unprepared for the Tule Narrows the first time I saw the place, but I've been back many times since and the feeling still hasn't gone away. From the top, on either side of the gorge, the first impression is of plains-scale Yosemite. Although it is wider downstream, in the Narrows section Tule Canyon pinches to less than one-half mile from rimrock to rimrock. Far, far below, just at 700 feet true vertical but twice that looking from a rim of the V, the stream threads a gorge with perpendicular red-brown sandstone walls that stand only about 75 feet apart. See it in the morning and the entire canyon is filled with a blue haze through which sunlight flashes off the stream. In the afternoon the air in the gorge is yellow, muting detail and color, and you realize, looking through enough of it, that you really can see air.

Tule's effect from inside the canyon, ascending it with no preconception of how it will look, was for me a strange mixture of delight, awe, and dissatisfaction. Reading my journal now, it is easy enough to pick out the emotional reaction. That last morning, camped in a turn in the canyon about two miles below the Narrows, we marveled at the remoteness, at how coyotes sat on their haunches 50 feet away and barked at us as if we were the first humans they had ever seen. Birds were everywhere, many of them species I had never seen on the Texas plains before. The water was clear, with more falls and pools and springs each mile. Cottonwoods were at peak autumn gold, the air still and balmy in the deep canyon. And ahead, so tantalizing that for three hours we dawdled and put off entering so the sensation would linger, was the gorge, one side in shadow, the other bathed in blue light.

Yet my journal shows a nagging dissatisfaction, not with the canyon but because of it. Toward the end I wrote, "There is not another state in the U.S. where a place like this would still be private land." Anyone who appreciates wilderness and has ever been in Tule Canyon will understand the sentiment.

NORTH OF LOS LINGOS CANYON, a descending line of mesas divides the waters of the Pease from the next series of streams in the Red River system. Knifing deeply through the plain across the next 25 miles are the narrowest and most strikingly colored canyons of the Southern Plains. Several—those cut by Mexican, Indian, and Coon creeks—no more than a mile or two long, are deep, plunging gorges that have a seep spring and a cottonwood grove or two and haven't been visited a half dozen times since the last band of Comanches camped in them. Defining this remote stretch of escarpment canyonlands at either end are three of the Texas plains' most beautiful canyons, sandstone marvels that preserve, in different ways, the wildest part of the Llano Estacado. Tule, named for the *Scirpus* bulrushes that line its creek, is the northernmost. Tule is a Nahuatl Indian word that at one time was correctly pronounced locally, but has now become Anglicized to sound like "tool." Its sisters are the flame red twin canyons of the Little Red River, since 1981 the site of Caprock Canyons State Park, one of the few parks in the Texas system developed under a modern plan to preserve a backcountry wilderness setting.

Wilderness is one of those words that's generationally diagnostic in a place like West Texas. For some it's a worship word, sacred. The very look of it can set a Sierra Clubber to scribbling off a check or a redneck environmentalist to start casting about through glazed and bloodshot eyes for survey stakes to pull. But for many, maybe most, rural West Texans beyond forty, wilderness is what their great-granddaddies fought and their granddaddies conquered in this country. Wilderness is the enemy.

The word is European, and it is revealing that most Native American languages are unable to translate it. Its etymology derives from the Anglo-Saxon *wildēor,* meaning of wild and dangerous beasts, which speaks volumes about the perceptional differences toward nature between Europeans and Native Americans. For nearly 2,000 years the western mind has been conditioned to the value-laden meaning of wilderness as it is used in the Bible, where it appears more than two-hundred times and designates a wasteland, a chaotic and dangerous antithesis to civilization. In the Old Testament especially, wilderness was synonymous with desert, a Middle Eastern pejorative that acquired renewed relevance when Euroamericans probed the American West. Biblical immediacy being what it is for both Catholics and Protestants on the Southern Plains, this old context for wilderness still has regional relevance. The pioneer legacy reinforces it. When old buffalo hunter and pioneer J. Wright Mooar told a plains audience that "any

PAGES 98–99:
Three views of the Yosemite of the Southern Plains, Tule Canyon—at 700 feet deep and only a quarter mile wide, the most dramatic wilderness of the Caprock Canyonlands

one of the many families killed and homes destroyed by the Indians would have been worth more to Texas and civilization than all of the millions of buffalo that ever roamed from the Pecos River . . . to the Platte," he was articulating a pioneer and Biblical value system about the wild.

Coexisting with that idea of wilderness is another one, also powerful in its influence. In the eighteenth and nineteenth centuries wilderness began to acquire some positive dimensions as a result of the Romantic movement, which recaptured something of the pagan feeling for nature and saw, instead of the "fallen" landscape to which humankind had been exiled in Genesis, a pristine creation direct from the hand of God, a sanctuary from instinct-deadening civilization where one might find religious inspiration and transcendent truths. The early geologist Sir Charles Lyell gave wilderness a new and understandable order in the 1830s; geology, in turn, provided a Rosetta stone for the breakthrough to Darwinian evolution. The famous historian of the American West, Frederick Jackson Turner, viewing societies through a Darwinian lens, added an intriguing twist: might not the American wilderness have shaped us as a people in the same way that natural habitats shape the evolutionary growth of any species? Texas historian Walter Prescott Webb would go on to argue that the Great Plains had created a Llanos American, who exploited technology like Colt revolvers and windmills and barbed wire to "adapt" to life on the plains. If all this were true, then wilderness was not—never had been—the enemy. More like creator.

These main elements (there were others) merged in the twentieth century into the wilderness emphasis of modern environmentalism. Environmentalism is not something that average West Texans are prepared for unless the protagonist is the federal government, as was the case when MX missiles and nuclear wastes threatened and aroused some regional resistance. They have been largely brainwashed by their politicians and their newspapers against other forms of environmentalism, which they tend to regard as either incomprehensible or as a bizarre form of tree worship. And, thank the good Lord, there aren't many trees on the plains.

The second largest state in the country has a burgeoning population of 16 million and until 1984, twenty years after the passage of the federal Wilderness Act, had acquired only a single designated wilderness. In that year the East Texas Wilderness Bill created five small wilderness areas out of national forest lands in the country north of the Big Thicket National Preserve. Texas' two major national parks, Big Bend and Guadalupe Mountains, are remote enough to offer some flavorful wilderness experiences most of the year. Pete Gunther, Justice William O. Douglas, John Bryant, Lloyd Bentsen, John Tower—environmentalism in Texas will remember you. But you're not even on second base.

TULE CANYON and the Caprock Canyons have historical associations, of course. People who read serious western history know of Tule, in particular, because of its role in the Red River War of the 1870s. It was in Tule Canyon that American troopers let loose one of the most naked expressions of hatred for all things Indian in nineteenth-century history. After surprising the southern tribes in Palo Duro Canyon and burning their villages and winter supplies, the troopers drove about 1,400 captured Indian ponies into a box canyon in upper Tule and slaughtered them.

Rancher Ray Adams, who possesses a face Richard Avedon should have photographed, took a small group of us—Jerry Griffin, an old friend from Austin; historian Fred Rathjen; Byron Price, director of the National Cowboy Hall of Fame; and myself—to the spot one eerie November day. Gauzy rain curtains, blown almost horizontally by the season's first blue norther, gave the unsettling impression that the day was about to fade into a time warp. Spooky, but the imagination wouldn't rest. Are those junipers, or troopers, crouching on the rimrock? That booming—is that the blood pounding from the climb? The north wind wheezes the last breath of a dying buffalo pony as it whistles around my collar. A raven croaks—it's real, I think, on that bare limb there—before blowing mist (or is it gun smoke?) obscures what in one frozen frame I almost took to be frenzied rearing and plunging.

We drove out of the canyon. Jerry and I stopped at the Texas historical marker on the highway. It was very matter-of-fact. It starts, "Two miles north of here Gen. Ranald S. Mackenzie, 4th U.S. cavalry, ordered shot the 1450 horses . . ." (Historical footnote: Ranald Mackenzie, who had never been known to accept the favors of women, openly and inexplicably began to consort with a prostitute within five years of this episode; he eventually was declared mentally unbalanced and dismissed from the service. He is now regarded to have suffered from the disease of the nervous system since named Viet Nam Syndrome.)

There were ancient trails through these canyons, and cowboy history, of course. But the real story of Tule and Caprock canyons is about wilderness. Simply, for three-quarters of a century and five scientific exploring expeditions, the headwater streams of the Red River eluded discovery, mapping, and understanding. The American government had been trying since Thomas Jefferson to find these canyons, but it was not until long after the Missouri, Arkansas, Colorado, and Columbia rivers were completely known that American geographers figured out the great Red River puzzle.

All during that long span of time when the Llanos were a part of New Mexico, the Spaniards knew and named all these places. Coronado quite likely was not here, but Pierre "Pedro" Vial definitely was in 1787. Tule already bore its name by then, and José Mares, another Spanish traveler with business between San Antonio and Santa Fe, gave the Caprock Canyons names that same year: he called South Prong Canyon, fittingly, Sangre de Cristo (blood of Christ) and North Prong Canyon, San José. Northern New Mexicans, Pueblos and Hispanics, had a regular trail to these canyons, which they traveled the way we drive interstates today.

Western science was far more confused. Employing a characteristically German syllogism, the Prussian naturalist Alexander von Humboldt settled the prevailing lacuna about plains hydrology by combining all the New Mexican plains rivers, the Pecos, Mora, Canadian, and

Red, into one river system in his map of 1804. Seventy years later government explorers were still trying to disentangle it all.

It took conceptualizing the Llano Estacado as a drainage divide, which seemed impossible, and accepting that the Red River began in draws and canyons, not mountains. The one good account of how it really was had been dismissed because it seemed to make no sense. But in the eighteenth century some unknown American trader, maybe it was Philip Nolan, provided the government one terse but accurate sentence. The Red River, he said, took "its source in the East side of a height, the top of which presents an open plain, so extensive as to require the Indians four days crossing it, and so destitute of water, as to oblige them to transport their drink in the preserved entrails of beasts of the Forest." And the name the traders gave the "height"? They called it the "High Plain."

"I THINK I've seen those sheep or goats you've been talking about."

A dark-haired young fellow was picking his way through the junipers and across a sandstone shelf in my direction. He sprawled on the smooth rock a few feet from where I sat, cross-legged on the edge of the precipice, and looked at me quizzically.

"There are some big animals below, by the water, and I think they're sheep or something," he said again, either for emphasis or because I hadn't replied. I nodded. Actually I was dubious. It was a warm and sunny winter midmorning and we had been, I thought, talking a little

In 1787 Spanish explorer José Mares named the two headwater canyons of the Little Red River Sangre de Cristo (Blood of Christ) and San José. Today they are LEFT, *South Prong and* RIGHT, *North Prong canyons of Caprock Canyons State Park.*

too freely and silhouetting ourselves too carelessly along the rimrock of the Tule Narrows to catch a band of aoudads unawares. Blake Morris is a fourth-generation West Texan whose grandparents still farm a piece of Llano near Blanco Canyon. Celtic genes bred true: he's strong, a good hiker with an Outward Bound course in his background. But he'd never spent much time in the deep canyons, nor seen wild aoudads. I figured he'd gotten a glimpse of something but probably mule deer. Not likely an aoudad.

It wasn't *an* aoudad. Being careful to skirt the clumps of hairy little starvation prickly pears, we elbowed to the brink of the plunge and looked 300 vertical feet past our noses. There, sprawled across boulders and in the grass around a noisy little waterfall were twenty roan-colored animals, several with flowing manes beneath their necks, and all except the scattering of juveniles with blackish, scimitar horns, more like goat horns than those of our native bighorn sheep. No mistaking, these were *Ammotragus lervia,* wild aoudads, or "Barbary sheep" as southwesterners are given to calling them.

For the next quarter hour we lay motionless on the lip of the Tule Narrows, watching an obviously healthy and contented band of animals that by all rights should not have been in this canyon. Aoudads, as indicated by the nineteenth-century exploring accounts of this area, are not native to the Red River canyons. In fact, they come from the Atlas Mountains of North Africa, where they are more correctly known as "sand goats." Calling the aoudad a "sheep" is something of a ploy; it is genetically more closely related to the goats than to *Ovis,* will interbreed with the former but not the latter. But what hunter wants to spend up to $2,000 to shoot a trophy sand goat?

They are wary, no question. Hunting has proved an ineffective way of controlling their populations, with a hunter success ratio of only about 25 percent. Watching this band, we got an excellent demonstration of why. Despite our near invisibility, some intuitive sense seemed to make them restless. When a coyote barked once from half a mile up-canyon the entire band was on its feet as if by one impulse. In single file, punctuated by many pauses as the leaders scanned for danger, twenty aoudads strung out across the canyon floor, bounding nimbly over boulders and into heavy cover. Within a minute they were a quarter mile away; in another minute they had vanished.

Seeing a band of aoudads in the wilds can generate excitement, particularly in those who don't really understand what the presence of this animal can mean to native wildlife in the Southwest. It is a big game animal (males can weigh up to 320 pounds) in a country where big game is now comparatively scarce. There are lots of them in these canyons now, by last aerial surveys about 1,800 in the Texas plains canyons and another 800 in the Canadian Gorge. But aoudads are literally walking time bombs of Old World pathogens. Those in the Texas canyons carry infectious bovine rhinotracheitis, two African species of lice, and more than seventeen species of intestinal worms, or helminths. They also compete directly with the native mule deer for forage, both animals preferring precisely the same species for browse and grazing but the aoudads demonstrating more plasticity in their diet. In other words, as one study puts it, because of aoudads "the survival of mule deer in Palo Duro Canyon may be affected seriously." The Bureau of Land Management has never won many plaudits from environmentalists, but its position toward aoudads is that their presence undermines the wilderness quality of lands they administer. As for the National Park Service, its policy has been put succinctly: "Quite simply stated, the species is uninvited, unwanted, and unwelcomed in the national parks."

Which introduces the obvious question: what is this animal doing in the Llanos canyons?

The answer is that, like most of the three dozen or so big game exotics introduced into the Southwest over the past half century, aoudads were released into the wild by landowners seeking revenue from hunting. Between 1949 and 1951 the Texas Parks and Wildlife Department had attempted to augment the small remnant mule deer population in the canyons by releasing 268 mule deer in Palo Duro, Tule, and Mulberry canyons. The reintroduction worked, just as it did in the Double Mountain Fork country. But it took almost two decades. Meanwhile, the hunters who dominated the New Mexico Department of Fish and Game in those years had released eight aoudads in the Canadian Gorge; by 1960 an aerial survey there counted 1,275 of them. Canyon landowners in Texas appealed for and got their own aoudad releases in 1957. And from an original population of 25 released into Mulberry, by 1964 there were 450; by 1978, 1,350; by the mid-1980s, 1,800. The aoudad is aggressive, resilient, and competitive. And it is just beginning its patterns of distribution and establishment of range in the Southern Plains canyonlands.

Every rule of sound ecological management was broken when the aoudad, without adequate study or exhaustive attempts at reintroducing the native big game, was turned loose on the plains. Now the only hope of controlling these animals is an ecological one. More and more of the wild bands seem to be contracting a disease called elaeophorosis, which causes large, scabby lesions ("battle scars" say hunting guides) that make the animals downright repulsive as trophies. It's the mule deer's revenge, caused by an endemic nematode that has no effect on the deer, and it has caused a 50 percent drop in aoudad populations in the Canadian Gorge.

Short of the reintroduction of wolves to the canyons, it's hard to think of a more satisfying wildlife development on the Southern Plains.

IT IS NOT, of course, that the aoudad is not a wild animal. It's that the concept of wilderness in its American form is so tightly intertwined with our history that the idealized western wilderness does not include introduced species, or any animals or plants that were not here when the first traders and explorers saw the country. Which raises the questions: what was here before the Euroamericans began their process of ecological imperialism and how much true American wilderness remains in these canyons?

We can guess at what was here, but there's a better way to establish an ecological baseline. A kind of time machine exists in the accounts of the first scientific exploring expeditions. They aren't perfect: they leave a good deal out, their species designations aren't always reliable, and their interpretive framework has to be seen as reflective of philosophies that are a century and a half out of date. But lacking a genuine time machine, they're the best we've got.

Lucky thing. Two U.S. Army topographical engineers' exploring expeditions penetrated the Red River canyonlands in the nineteenth century, searching for the elusive headwaters of the Red. Two, because the first one in 1852 proffered a fascinating but confusing report so that a second, until recently forgotten, expedition had to be dispatched in 1876 to supply vital final pieces in the geographic puzzle. The earlier one was the Marcy Expedition, led by Randolph Marcy and George B. McClellan with geologist-naturalist George Shumard along. The second was the Ruffner Expedition, commanded by Ernest H. Ruffner, guided by buffalo hunter Billy Dixon, and including Charles A. H. McCauley as ornithologist-naturalist. They left us journal descriptions, natural history lists and illustrations, a window into time. Both were conducted at favorable seasons for natural history, both while the canyonlands wilderness was still intact. Lucky thing.

The American scientific explorer has been called an anthropological type, a gatherer of knowledge who was programmed by his civilization to satisfy a whole range of cultural desires with his discoveries. By the time of the Marcy Expedition the "type" was well educated, affected to greater or lesser degree by the Romantic inclination to view wild nature as a grand thing, and almost to a man believers in Creationism and the Great Chain of Being—in other words, believers that the creatures and even the landscapes they were discovering had existed unchanged since the time of Genesis. As a result, the raw data sometimes seem an incoherent mass. Modern ecological relationships often have to be inferred. Gregor Mendel, working with peas in his little garden in Austria, and Charles Darwin, exploring a single island chain in the Pacific, gleaned greater insights into how nature and life work than all the early western explorers combined.

What we know now, for example, is that the vegetation and the wildlife found in Tule and Caprock canyons in the 1800s were products of a complex and long history and many diverse interactions between soil, space, and light. We know also (and this, at least, the explorers were able to see dimly) that these canyons were far richer in topographic and, hence, species diversity than any of the surrounding lands for hundreds of miles, that they are so located that their native plants and animals comprise a unique blend of eastern and western types whose range perimeters just overlap here. An additional century of nature study explains relict species and specially evolved endemics to us. Marcy, Shumard, and McCauley stumbled when they met them.

Randolph Marcy was a Massachusetts Yankee who, so his biographer Eugene Hollon claims, ought "to be ranked alongside John Wesley Powell" as a western explorer. Powell, those familiar with western history might recall, was the fellow who first descended the Colorado River, then did it again, later founded and became director of the U.S. Geological Survey, founded and directed the Bureau of Ethnology, the Western Irrigation Survey, and so on. Marcy's greatest accomplishment is said to be his discovery of the origin of the Red River, his most famous literary passage his account of the final ascent to its "head spring." He is a regional hero. He also seems to have perpetrated a gigantic hoax that he somehow persuaded the members of his party not to divulge.

A century and a half later, the secret is out: Marcy never found the head of the Red River. In fact, Marcy didn't get within 125 miles of the Red's origin. He didn't even explore the right canyon, abandoning Palo Duro down at its mouth, probably because it was dry or the water was too gypy in late June. Instead he spent three days ascending Tule Canyon as far as the Narrows, called the spring there the origin of the Red, and then tried to tell readers that the canyon ended in a box where "the gigantic escarpments of sandstone, rising to the giddy height of eight hundred feet upon each side" united overhead. A dozen historians have struggled trying to find such a spot in Palo Duro Canyon. It's not there. Instead, the Marcy lithograph of this scene is a dead ringer for the wall that towers above the main spring in the Tule Narrows, except that in reality the canyon makes a right turn and continues on for several more miles.

But the point is not so much that pudgy, mutton-chopped Randolph Marcy made his fame (and the expedition did make him famous) through flimflammery, but that much of his and Shumard's natural history is 1852 Tule Canyon natural history. It was Tule, not Palo Duro, that Marcy was describing when he wrote that "the magnificence of . . . these grand and novel pictures . . . [of] unreclaimed sublimity and wilderness . . . exceeded anything I have ever beheld."

The Marcy-Shumard reports gives us a generalized overview of wild Tule Canyon biology. No birds were collected or described, but Marcy himself listed twenty-four mammals, dominant among them bison, which were said to winter deep in the canyon, pronghorn, black bears in large numbers, and a cougar that responded to a deer bleat. They saw no elk and, very oddly, reported only white-tailed or "common" deer, even though the mule deer had been known on the Llanos since Titian Peale of the Long Expedition had painted one from the upper Canadian. They saw no grizzlies, either. David Brown, in his recent book, *The Grizzly in the Southwest,* doesn't think there were grizzly bears on the Southern Plains. But neither Brown nor Marcy was aware that in 1826 a trapper-surveyor named Alexander LeGrand twice reported having seen "white bears" along the escarpment of the Llanos.

The herpetology collection from the expedition was considered one of the most significant to come out of western exploration; along with the Mexican Boundary Survey it revealed reptile and amphibian richness in the Southwest. Ten species of snakes (three of them thought to be new to science turned out to be already known), six of lizards, a toad (a Western spadefoot), and a frog (a Great Plains narrow-mouth) were unlucky enough to get sacrificed to science. Two common canyonlands creatures, the imposing but inoffensive Texas brown tarantula (*Dugesiella hentzii*) and a myriapod, the giant red centipede (*Scolopendra heros*), were unknown species. The botany lists catalog most of the major trees and shrubs one would expect, among them honey mesquite, plains cottonwoods, black willows, hackberries, and sumacs, and added the wild china tree, skunk-bush sumac, and a new *Artemisia* (sand sage) to Linnaean science. The explorers noted but misidentified the Mohr oak, which can be both shinnery and large trees, and where four junipers grow they listed only one, a species not found in Tule Canyon. Nothing about ecological relationships. Just lists. It was a science of pigeonholing, not grand theory. But it's something to go on.

In the quarter century after Marcy, certain folks began to suspect chicanery. Army commanders chasing Indians kept encountering draws atop the Llano Estacado that pointed toward the Red River. In the spring of 1876 Lieutenants Ruffner and McCauley had their try at the canyonlands wilderness. They did considerably better, not only because Ruffner was a more honest explorer than Marcy but also because McCauley was an energetic and first-rate ornithologist. And it's the bird work, more than any other of these early natural history accounts, that provides an environmental baseline for gauging change over the intervening century.

The Ruffner expedition set up base camp near the falls in Palo Duro in late May and spent the next five weeks exploring all the surrounding Red River canyonlands. Billy Dixon personally guided Ruffner into the Tule Narrows. The West Point officer reacted about as Dixon thought he would, as all of us who have come upon the Narrows unprepared have: he was flabbergasted and delighted, later writing that it was the "narrowest and deepest cañon met by us in a month's steady search of them. . . . Where [the creek] broke into the cañon I saw the finest exhibition met with during the trip of the perpendicular walled cañon. . . . The barometer showed a fall of 500 feet." And it was rank with verdant vegetation, "a natural abiding place for birds."

But such birds! Where today one sees mostly western warblers like the orange-crowned and Wilson's warblers, McCauley found warblers now associated with the eastern forests—prothonotary and yellow-throated warblers—nesting in the canyonlands. He saw a wood stork, the only known West Texas sighting of this bird, rise from a pool in Palo Duro and a crested caracara feeding on bison carcasses in Mulberry Canyon. Wild turkeys were incredibly common (by 1905 they had been extirpated); so were black vultures (no longer found here) and both common and Chihuahuan ravens (now extremely rare). Long-tailed chats nested in the dense vegetation along the canyon streams, and prairie falcons and pyrrhuloxias in South Cita Canyon. Neither bird nests in the canyonlands now. Four other canyon nesters of that time are now on the "early warning" Blue List of threatened birds: Swainson's hawks, nighthawks, yellow-billed cuckoos, and warbling vireos.

McCauley saw bobwhites, but no blue quail in 1876, even though he looked for them. He noticed a few ladder-backed woodpeckers, but nothing like the numbers that have come in with the mesquite and yucca increase. He did not see a single scrub jay or cactus wren or curve-billed thrasher, although all three are in Tule and Caprock canyons now, the scrub jays impossible to ignore. He did not see a single Mississippi kite. It must be true that their range shifted westward when the 1930s shelterbelts were planted.

McCauley's bird observations in the Red River canyonlands, when compared with bird lists from a century later, document a rather remarkable transformation of the original wilderness. The trend clearly has been toward desertification, brought on not by climate change but by human changes, even deep in these remote canyons. The enormous drawdown of the aquifer has sucked the power from the springs. Lush, riparian willow–big bluestem–cottonwood communities have shrunk to the point that they are now a threatened community in Texas. The giant Rocky Mountain junipers have almost been entirely removed from Tule and Caprock canyons. Pioneer loggers, using horse-powered winches and cables hundreds of feet long, are supposed to have cut enough out of Ross Canyon, a side gorge of Tule, to build a telegraph line from Wichita Falls to Abilene. Salt cedars, pigeons, aoudads are here now; bison, bears, wolves are not. The slaughter of plains wolves probably contributed to the disappearance of many of the original nesting bird species here, as raccoons, skunks, and weasels, once held in check by wolves, increased. And the desert effect, like some sere brown shadow lengthening northward with the twilight, keeps spreading its influence and its species as I write this.

We know a great deal more now than the explorers could have suspected about how this extraordinary, almost-desert, West-meets-East ecology was created. The essential vegetation of the Llanos canyonlands evolved in the Southwest about four million years ago, when local mountain rain shadows created habitats that selected for small-leaved and drought-resistant plants. This so-called Madro-Tertiary Flora spread northeastward to contribute characteristic canyonlands species like hackberries, the delicate little forestiera trees, sumacs, Mohr oaks, wild china, algerita, and mountain mahogany, all of which became well established in the middle moisture ranges of the canyons.

During the Pleistocene glaciations, Rocky Mountain species—blue spruce, Douglas fir, ancestors of the ponderosa and piñon pines, one-seed and Rocky Mountain junipers—colonized eastward into the draws and canyons. But the Altithermal drought drove the spruce and firs off the plains, shrank the pine range westward, although both piñons and ponderosas remained in the Canadian Gorge, and left the two western junipers as relict populations occupying the more mesic (wet) sites in the canyons. The dry pulsations also sent mesquite and cactus colonies north from the subtropical scrub-brush communities, and they naturally took to the xeric (dry) slopes. Some of the abandoned communities have been here long enough to evolve endemic forms. Piñon mice were left with the mountain junipers; they've evolved into a distinct type, the Palo Duro mouse (*Peromyscus truei comanche*), whose nearest relatives are 300 miles away. The red-berried juniper of southern Arizona and the red-berried Pinchot juniper of the canyonlands have differentiated in isolation from one another, although once they were the same. Now it appears that Pinchot and one-seed junipers are hybridizing. Eventually they will evolve an endemic canyonlands juniper.

These same climatic cycles created the mixed prairie grassland complex in the canyons. During wet millennia the bluestems colonized westward from their evolutionary home in the Carolinas. The cool-season grasses, western wheatgrass, the Stipas like needle-and-thread, were left when glacial pulses brought them south. The warm-season maturers, the three-awns, side-oats grama, blue and black grama, and buffalo grass, are all contributions from the Quaternary droughts.

This kind of rich multiplicity was possible in the canyons because of the extraordinary range of soils, moisture regimes, and shade conditions and because geology proceeds so rapidly here. Slumps, rockfalls, landslides produce unending disturbances, perfect for herbs that thrive in subclimax conditions. And deep canyons, steep slopes, cool north exposures, sunny dry ridges are classic locations for refuge species.

Spend time observing in the canyonlands and patterns emerge. Here is mesquite, on the valley floors and lower slopes, but not higher up. Why? Is it elevation, or something else?

It's something else. There is not enough elevation difference from the canyon floors to the rimrock to change the dominant species. Exposure and the soil effect that botanists call "particle size", which in turn affect moisture retention, are the critical elements in the pattern. Mesquite has a wide tolerance for particle size and soil type but a low slope tolerance; because of its lateral support roots, it prefers level ground. Pinchot juniper has a wide tolerance on an exposure scale but a low one in particle size. It is found on slopes because it extracts the moisture protected beneath boulders. One-seed junipers do the same but prefer a steeper, cooler situation. The two sumacs mirror the same pattern, with small-leafed sumac colonizing dry north slopes and skunk-bush sumac preferring shady south ones. Rocky Mountain junipers once must have been near the springs and under the west and south rims of most of the Texas canyons, the coolest, wettest spots east of the Rockies.

These are the kinds of things anyone interested in wild nature in the canyonlands ought to take pains to notice.

Chris White and I descend a crack in the Tule cliffs to an abandoned golden eagle's nest. Eagles are making a comeback in these canyons.

ON A STILL, beautiful July morning in Tule Canyon, canyon wrens flit through the rocks a few feet away, trilling their haunting melody of the Southwest. Watching them, listening, I repeat Roger Tory Peterson's wonderful canyon-wren line, "voice: a gushing cadence of clear, curved notes tripping down the scale."

Pigeons soar out of the shade and flash into sunlight. The sound of little rapids in the creek 250 feet down rises and fades with unfelt air currents. During the ebb intervals I can hear the pounding of my pulse and the plopping of sweat dripping off my nose onto red sandstone rock. I shift and pebbles cascade into the blue haze, creek bound in a Dopplerian disappearing act. That 250 feet is what you call your clear—good-bye!—vertical plunge.

I am wrapped around a boulder that blocks an already slender enough 15-inch ledge, waiting for Chris White to scoot sideways on his bottom, his legs dangling in the blue, far enough to make room for me to do something other than hang in space. Chris is chattering happily about the spot, the morning, the birds. He's been here before, has done the rather interesting little 25-foot technical climb down a cliff crack to reach this ledge dozens of times. By the time I finally got my fanny firmly on the sliver of terra firma, he was out of sight, exclaiming about a cave and other nests.

It was nests that had brought us down the vertical red slickrock of Tule Canyon, two untidy scrambles of branches and grass and rodent braincases. Even an inert egg, which we discovered when it imploded in a sulphur mist as I crawled across one of the nests. Chris, a wonderfully talented wood sculptor who works with big juniper logs he hauls out of Tule and uses an old ranch house on the rim as a studio, has been watching these and other nests in the canyon for years.

One noteworthy wilderness apotheosis that was in Tule and Caprock canyons when the explorers first saw them is still here. Eagles.

Eagles. Majestic, large, a soaring silhouette that ranks up there with bugling elk and the exploding popcorn effect of running pronghorn as the essence of American nature. Despite population ups and downs, eagles are still here. When the aquifer was higher and the springs gushed, both bald and golden eagles nested in these canyons. Explorers and early canyonlands visitors mention big congregations. Too big to suit the old-time ranchers and their cowboys, who shot and poisoned them by the thousands and, with airplanes, launched an infamous vendetta that slaughtered an estimated 20,000 golden eagles in West Texas after World War II, earning national admiration for modern ranch-

The sandstone "mitten" of South Prong Canyon, Caprock Canyons State Park

ers in Trans-Pecos Texas. Bald eagles ceased to nest in the canyonlands in the early twentieth century, although they continued to overwinter in Palo Duro in large numbers (National Park Service investigators saw more than 100 in four days in 1938). Their numbers plummeted again during the years when DDT was sprayed across the plains croplands, but bald eagles are back in upper Palo Duro in good numbers most winters now.

Golden eagles, more prone to favor rocky canyon country, have been gradually rebuilding their population in the canyonlands since DDT was banned and ecologists proved that stock losses to eagles was a sham. In the 1950s ornithologists knew of nearly thirty nesting pairs up and down the Caprock. That number shrank to seven during the decades of unrestricted pesticide use. Now, despite their susceptibility to Compound 1080, the once-banned coyote predacide again in use in the West, the Llanos eagle population is slowly rebuilding. At least three pairs nest in the Double Mountain Fork Canyon, one in Blanco, another in Los Lingos, and at least one pair in Caprock Canyons. Two pairs, certainly, probably three or four, raise broods in Tule, and several nesting birds are in Palo Duro. In five years of looking I finally saw a pair investigating Yellow House in the early spring of 1988.

Despite all the changes, these canyons yet retain that indefinable quality that whispers *wild.* The eagles, the few remnant mule deer, the oddly tall Rocky Mountain junipers in Tule's Teepee Pole Canyon and by the upper pool in the depths of North Prong Canyon, the shimmering gallery of giant old cottonwoods in lower Tule, coyotes as territorial as dogs as you pack into the remote inner recesses: all these can be seen as remnant, but also as foreshadowing. These canyons have held on. Two have now begun the process of rebuilding. We badly need the third officially set aside as part of that process.

"I HAVE A SUGGESTION and you know what it is."

I am calf-deep in a clump of blooming Indian blankets and sleepy daisies, witnessing life through a camera lens. It is May in Caprock Canyons State Park. The gaillardias and Tahoka daisies spread bright Navajo rug coloring beneath mesquite trees heavy with plump, caterpillarlike, yellow flowers. In every direction against the red sandstone backdrop, yucca flower pods gleam a creamy greenish white above the grass. It's that one time in the otherwise dullish yearly cycle of the native soapweed. Sexuality for a yucca is not, as it is for us humans, a full-time job. Reserved as they look, these yuccas are in a bacchanal, pollinating one another in wanton frenzy.

I straighten up and look around. Vertical. Red as blood (José Mares, you were right). Deserty (a friend's five-year-old daughter: "Daddy, this isn't a park. It's a desert"). In this light it looks like . . . like the set of a roadrunner-coyote cartoon.

"Don't you think you should pay some attention to me? I've thought of something we should do, and this is the morning for it."

My eyes are still blurry from microfocus. But a vision is standing in the trail, hands on hips. The vision is decidedly gender specific. The sun is not too high yet and it's behind her, illuminating a soft white down covering her forearms, her thighs. Eyes exactly the blue of the dayflowers we've seen crackle with vitality under a halo of blonde that frames her face and spills past freckled shoulders. The hell with flowers.

"A hundred years ago only a fool or someone who enjoyed fighting would have brought you into this canyon. You'd have ended up over the saddle of a Kiowa or Comanche war pony and it would have taken an awful big pile of goods to ransom you back in Taos or Santa Fe." *Careful.*

Her eyes were dancing, mocking. "You did bring me here then, don't you remember?"

We turn up the trail, heading west and upcanyon. Barely an hour earlier we'd left my truck at the "Scenic Drive" turnaround and entered the portals of South Prong Canyon on foot. About 125,000 people a year now visit this park, at 13,906 acres fourth in size in the developing Texas state park system. Most bring RVs and televisions and stay in the campgrounds near the splashing, angling potential of Lake Theo, back at the park entrance. A small percentage try the trail up South Prong Canyon, make the short but rough scramble over Haynes Ridge and into the North Prong Canyon, then descend along the boulder-strewn drainage past fern caves and pools and red walls, a hiking loop of about eight miles, which is what's wrong with it. There are another nine miles of backpacking-equestrian trails in the park, two primitive campgrounds if you don't count the ones the farm-worker illegals use. The Texas Parks and Wildlife Department has come a long way since it wanted to build swimming pools and tennis courts in Palo Duro, and they've essentially left the canyons of the Little Red River alone. Beautiful and wild as it is, there's just not enough of this park, not enough to really lose yourself in for several days or a week.

But it is a Wednesday and Caprock Canyons is not only a new park, it is also 75 miles from the nearest city of any size, in one of the most sparsely populated counties (Briscoe) in the state of Texas. So we have the entire backcountry to ourselves and the mockers are trilling and the scrub jays are flitting and squawking and once, I tell her, I woke up right here and golden eagles were flying the cliffs the way they do in the morning and she'd better be careful with her suggestions on a day like this in this canyon.

A quarter mile past the primitive camp in its grove of cottonwoods and the 50-foot red sandstone formation I call The Hand (a sort of junior version of one of the Monument Valley Mittens), the trail pitches upward. Aoudads clatter rocks in the stillness. I go first because the trail is sometimes hard to see and I point out a little geology, Permian here, Triassic there, 200 million years for this rock and her ancestors were little shrews scurrying from big lizards then. Saying this I turn to look at her and feel a warm glow for evolution.

She is the daughter of concrete canyons like Newark and New Orleans and now Houston, come to West Texas with the universal impression that the plains are flat and uninteresting and serve only to hold the rest of the world together. But she's been a dancer and however unexpected this up-and-down slickrock country is, this trail is no sweat for her. So when we reach a place halfway up, where runoff has cupped out an amphitheater in smooth terra-cotta rock, I'm the one who calls for a moment of meditation on the view. Sprawled in the shade on the cool sandstone, swilling tepid grapefruit juice, I realize I'm in trouble and it's not the climb. The sunlight is brilliant and the goddamned mockingbirds won't shut up and I can't keep my eyes off Robin.

The feeling seems emphatically mutual. Out in the bright gorge a prairie falcon dives at the pigeons. Just above us green junipers creak as a light breeze gets up. There is absolutely no one around and no real pagans could resist. In less than a minute we are naked and laughing and boldly admiring one another. Ah, Pan. Ahh, Joe Meek and Jim Bridger. Yes, all you unnamed-but-not-forgotten waifs and nymphs and spirits of forest and cave. Then she pulls me down against her, and the dirt is red and the rock is red and her eyes are blue springs in the desert as we reaffirm that praiseworthy anthropoid capability for lovemaking on a surface of virtually any contour that occurs in nature.

Afterward I am dangling my feet off the ledge of a pour off and soaking up vitamin D. She sits down beside me, hair shining, smelling clean and a little sweaty and wonderfully female. I take a piece of Permian clay I picked up in the stream and with my thumb lay a terra-cotta streak across her forehead, another on her thigh.

"Have you ever read Edward Abbey's essay, 'Freedom and Wilderness'?" I ask her. She hasn't. She knows Abbey, but only from *Desert Solitaire.*

Too bad, for the sake of the American wilderness idea, she or someone like her didn't meet Thoreau. We'd have never had to wait for Abbey for a literary salute to the sensual dimensions of the free and the wild.

. . . about 11 A.M. all at once we, that is Teodoso and myself, stood at the brink of the cañon of the Red River, and it was the grandest sight I [have] had in this great United States.

—*Illustrator* **Carl Hunnius,** *1876*

It is absurd how much I love this country.

—Georgia O'Keeffe, *1916*

Landscape topography as historical evidence: profile of the cliffs above the spring in the Tule Narrows leaves little doubt that the 1852 Marcy Expedition lithograph View of Head Spring of . . . Red River *was of this spot.*

Chapter 6

Visions of Palo Duro and Other Canyons of the Imagination

EXAS GOVERNOR Pat Neff seems to have thought he was conferring high praise when he likened it to Colorado. To Joseph Wood Krutch it was an advance taste of New Mexico, and for Georgia O'Keeffe "not like anything I had known before." But the most revealing written impression of Palo Duro Canyon is still George Wilkins Kendall's, set down a century and a half ago, when the Texas Republic's filibustering expedition against New Mexico was disintegrating in the alien, high desert 400 miles northwest of Austin. Kendall was a journalist in an advance party being escorted back across the Llano Estacado by one of the Comancheros the Texans had come to liberate, but even that humiliation couldn't dampen his reaction when the smooth ground beneath them abruptly plunged away into Palo Duro. It stunned them with "a spectacle, exceeding in grandeur anything we had previously beheld," he wrote. As an easterner, there was only one sight, America's most famous Romantic Age scenery, he could compare this to. "Regularity was strangely mixed with disorder and ruin, and nature had done it all. Niagara has been considered one of her wildest freaks, but Niagara sinks into insignificance when compared with the wild grandeur of this awful chasm."

It's Kendall's word choice that I like, notwithstanding the occasional failure of his cliché filter. "Wild," "grandeur," and "awful" were 1840s buzzwords and they center Kendall in a time as effectively as if he'd scribbled "Far out!" in his journal. But "disorder," "ruin," "freak," and "chasm" bespeak something different, a mind grappling with how to define the heretofore visually unimagined.

Kendall, I have come to think, may have been trying to tighten an octagonal bolt with a hexagonal wrench. Without a lengthy resort to geological explication—and neither the writer nor the science was up to the task—how could Kendall really convey a canyon scene to an audience that could not imagine such a thing? The fact is, he couldn't. But a visual artist, a painter, could.

CULTURALLY IT'S EXPLAINABLE but still disappointing that bold, bright, surprising Palo Duro Canyon has not inspired more musical or literary art, Michael Murphey's video tribute excepted. West Texas has a well-known musical tradition, but while wind has been a persistent theme of native songwriters Bob Wills, Jimmie Gilmore, Butch Hancock, Terry Allen, and Joe Ely, canyons haven't been. Despite its status as a regional event, Palo Duro's summer musical, *Texas,* is nothing much else than a soapy pioneer tribute that has failed to produce any Rodgers and Hammerstein–quality songs with a life of their own; its treatment of the native peoples is patronizing and historically ridiculous. As for literature, aside from Phebe K. Warner's 1930s newspaper articles, written to support the creation of a million-acre "National Park of the Plains" during National Park Service hearings, nothing approaching the caliber of John Muir's *The Mountains of California* has ever been written about Palo Duro. Muir himself seems not to have ever seen the canyons of the Southern Plains, although Enos Mills, the "Muir of the Rockies," saw some of Palo Duro and made a speech in Amarillo calling for a park.

In contrast, visual artists have had a great deal to say about the Llanos canyonlands, Palo Duro in particular. Why painters and not nature writers or musicians? I think it's because Palo Duro is so overwhelmingly a visual experience, of forms and colors and light, that the purely visual resonates far more powerfully than history or mythology or biology, or even geology, do. Topophilia for this canyon has surged most strongly through the tuned antennae of the visual artists. Landscape art and landscape artists loom prominently in Palo Duro's history of place, and the artistic images act like cultural barometers, across time, of how Palo Duro has exuded a landscape energy that affects humans. How such energy from landscape gets refracted through the lens of personal vision and prevailing culture is the fascinating part.

PALO DURO CANYON is one of the two largest canyons on the Southern Plains, is the most exciting visually, and is the only one of the Llanos canyons most people have heard of, a sort of Grand Canyon of the plains. I saw it for the first time in 1978, the way most people do, by driving into the state park and gawking incredulously at the cliffs and banded badlands from a car window. This is, according to landscape historian John Brinckerhoff Jackson, a way of experiencing nature that we may as well get used to, for this is the way most of us see the countryside these days. Ten years later my own involvement with Palo Duro is more varied and intimate, but I'd still rather see Palo Duro from a car than not at all.

Since this is the usual way to come to know Palo Duro Canyon, for the most authentic feeling from what was once actually called the Grand Canyon of the Llanos, for a feeling that might give you a good taste of what the nineteenth century Romantics used to call "sublime," drive sometime across Palo Duro on U.S. Highway 207, preferably at one end or the other of the day, when the light is rich and slanting. This is called the Claude Crossing, or Paradise Valley, and like the Hanks Crossing of another conglomerate region on the Colorado River in Utah, not so long ago it was a dirt two-track. Also like Hanks, and unlike the state park upcanyon, this part of Palo Duro is wild and undeveloped, with few people in a wide-open country that is a long way from anywhere else.

The ensuing spectacle—Palo Duro is at this point a 12-mile-wide, 800-foot-deep roar of color, a jumbled and jarring antithesis to the linearity of the plain and the brooding silence of the overhanging sky—must have been working various kinds of magic on the human psyche for centuries.

Leave your camera in the car and stand at either rim. Take in the canyon with your senses. Grip a juniper, inhale the scent, feast the eyes on reds and greens. No one has to know the first thing about botany or geology (it's red, it's layered, it's a broken puzzle) to appreciate this scene. There are ancient reasons why we react emotionally to a landscape like Palo Duro, ones that cross cultural lines and are partly prehuman. It's the colors. Color vision may have evolved among primates so that ripened fruit could be picked out against the wall of green vegetation. A vivid red-and-green landscape perhaps triggers within our brains primal reactions of satisfaction or excitement. On this response are layered cultural ones of great variety and value: religious feelings of communion with nature, literary, mythical, or historical allusions, awareness of ecological diversity. Thoughts like *freedom, wildness, good hunting,* or perhaps *deserted wasteland* are associated with scenes like this but are obviously cultural constructs. At bottom, as a Sioux hitchhiker who once traveled across the West with me put it, pretty eloquently I thought, Palo Duro is simply "good to look at."

CANYON, TEXAS, in 1916 was not the likeliest town in the Southwest to serve as a base for an inspired new sort of landscape painting. Taos, 300 miles to the west, by contrast, already looked as if it had aspirations. Ernest Blumenschein and Victor Higgins had left New York for Taos, and other painters were coming, more every summer. East of Canyon about the same distance was Dallas, which had Frank Reaugh, who since 1889 had been making long sketching tours to the nearest western landscapes with talented students in tow. But in 1916 a serious, 28-year-old midwesterner stepped off the train in Amarillo excited at the prospect of a return to the Southern Plains and about her new position teaching drawing and painting at the teachers' college in Canyon. Within two years the interaction between the Llano Estacado and the maturing artist, poised in her growth for exactly such a stimulus, would produce a solo New York exhibit that stunned viewers with pure, simple landforms and hot colors taken from the heartland of the continent. The artist not only gave Americans of the Great War era a new vision of nature, she also gave them a new landscape to visualize.

Overlook of Palo Duro Canyon

Wilson Hurley's From the West Rim (Palo Duro Cañon). *Courtesy La Plante Gallery, Albuquerque, New Mexico.*

Georgia O'Keeffe—not the somewhat gruff old lady of the New Mexico years but the slender, dark-haired young woman of Alfred Stieglitz's photographs—was that artist. And what she found in the Texas Panhandle was the American Southwest in all its sweep and atmosphere, simplicity of line and visceral power. O'Keeffe's star is the brightest, but she is not the only painter to have been inspired by the plains skies and canyons. To really understand the Llanos canyonlands and the evolving interaction between place and art, one must work both forward and backward in time from the O'Keeffe years.

Landscape painting in Europe and the United States in the years between about 1825 and 1890 was dominated by Romanticism, in the case of the United States, in fact, by what art historian Barbara Novak has called a "culture of nature." Initially, American tastes in topography and the painterly rendering of it were shaped by the Hudson River School, whose disciples sought out and portrayed the natural beauty of the eastern mountains in quests that were literally religious pilgrimages. Western art at the time meant mostly Indian portraits and scenes done by artists like George Catlin, Karl Bodmer, and Alfred Jacob Miller, since Indians and the Republic of Mexico controlled most of the interior West.

In the same years that Hudson River painters and transcendentalist writers were exploring nature in the East, the second generation of American explorers was getting its first taste of the Southwest. The initial reaction can be summarized in a word: alienation. While the Rocky Mountains evoked comparisons with the Alleghenies and the Blue Ridge, they seemed islands in a strange and forbidding desert sea. The first American explorers to cross the Llanos—on the Long Expedition of 1819–1820—returned with the Great American Desert impression that colored American ideas about the area for almost a century. Long was writing directly about the country along the Canadian River, which he explored with a contingent which included landscape painter Titian Peale. From a famous family of painters and naturalists, Peale evidently found the barren plains country so unlike contemporary ideas of natural beauty that, except as backgrounds for his paintings of mule deer and burrowing owls, he ignored its possibilities for landscape.

Alienation bordering on revulsion continued to be the reaction of most Americans to the Southern Plains for at least three more decades. During the 1840s, before U.S. aggression swept away Mexico's title to the land, other explorations penetrated the Llanos along the line of the Canadian River, collected natural history specimens, and speculated on the geology of the plains. But they left little in the way of painterly representations of the plains. And as yet no one with a palette or a sketch pad had seen Palo Duro and the canyonlands on the upper Red River.

This was true, of course, of such soon-to-be famous western canyons as Yosemite, the Grand Canyon, and the Canyon of the Yellowstone, too. Aesthetic tastes in New World scenery stood on the threshold of change in the middle nineteenth century, as Clarence Dutton, in his perceptive 1870s exposition on how to appreciate canyon scenery, well recognized. From the time of the great railroad surveys in the

1850s through the reconnaissances of George M. Wheeler, Ferdinand V. Hayden, and John Wesley Powell in the 1870s, painters and topographical illustrators, along with the use of inexpensive lithographic reproductions, would make the alien landscapes of the West familiar and compelling as American scenes.

Landscape-art interpretation routinely lumps the canyon scenes painted then with the landscape revolution of the Rocky Mountain School, pioneered by painter-adventurers Albert Bierstadt and Thomas Moran. But the apparent responses of American culture to paintings of the mountain West (which is, after all, similar to, albeit grander than, mountain scenery along the Hudson) and to those of the western canyonlands country were different, it seems to me. Today the two types of landscapes are not of the same school at all, defining very separate regions and states of mind. Canyons in their dry, desert form were strikingly weirder to nineteenth-century tastes than mountains. The barren slopes, speckly vegetation, and often vividly colored rock; the blockily vertical tiers exposed by erosion; and, especially, the eerie psychological effect of entry into the earth (which makes, as I've elsewhere mentioned, a canyon ascent very different from a mountain climb) have always meant that canyons exude a distinct kind of energy.

And I was excited when I realized, after a little work piecing together the chronology of exploration publications in the 1850s, that the first southwestern canyon landscapes the American public saw were of Palo Duro Canyon.

In 1854, before the first of the Pacific railroad survey reports was published with the initial representations of the Grand Canyon, done by F. W. Von Eggloffstein, Randolph Marcy's 1852 report of his exploration of the head of the Red River was ordered printed for the public. Several thousand copies at one dollar per copy were sold. It came bound with seven landscape lithographs, which gave a public then as interested in western landscape as we are in popular music, and almost as inclined to make culture heroes of those involved, a taste of the Southwest.

Considered together, these lithographs of the Red River canyonlands dramatize an interesting reaction by Americans of the day to canyon landforms. In his *Man and the Landscape,* Paul Shepard has advanced the idea that "there is no better example of the evocative power of natural landscapes than the response of westering pioneers to novel erosional remnants and angular cliffs." These lithographs from the Marcy report, done by Ackermann Printers of New York City from Marcy's descriptions and geologist George Shumard's sketches, may be the best illustrations for making Shepard's point that we humans see our environment through the filters of our time and culture.

What we saw in the 1850s were castles, cathedrals, and the skeletons of classical architecture in canyon scenes. The usual explanation is that Americans believed they had no historical icons comparable to those of the Old World and, instead, substituted grand landscapes wherein they professed to see ruins and other visible symbols of antiquity. While true on one level, this still doesn't really explain why European painters like Karl Bodmer and Charles Preuss inferred similar images from butte, mesa, and rimrock formations. In fact, pioneer literature from both Europeans and Americans is full of ruin impressions from erosional western landscapes. Clearly this was a European response, as much an attempt to make the alien somehow familiar as the application of familiar names (i.e., *mesa, butte*) had been in an earlier time.

These first Marcy lithographs of the Llanos country tell us rather less about the plains canyons than about the way Euroamericans perceived canyonlands country. *Border of El Llano Estacado* showed a sheer, mountainous terrain on the Southern Plains, the effect of the escarpment's height accentuated by towering cumulus clouds. A Romantic Age convention—contemplative human figures evaluating the scenery while stiffly decked out in formal attire—provides scale and a point of entry for observers. Midway up the wall of the central peak is a rimrock formation, which the artist has represented as a ruin. A similar ruin motif figures prominently in *View Near Head of Red-River,* a scene that comes geologically closer to showing the staircase tiers of the Llano canyons, and it dominates *View Near Gypsum Bluffs on Red River,* which eschews all subtlety and reveals the presence of a naturally formed Greek temple with columns supporting a classical dome. The final canyonscape, *View of Head Spring of Ke-che-ah-que-ho-no, or Red River,* is visually the most impressive of the set in spite of the fact that there is nothing remotely like it in Palo Duro Canyon. This lithograph is the smoking gun on Marcy, the final evidence that he never explored beyond the mouth of Palo Duro, for *View of Head Spring* can only represent one place in the Red River canyonlands: the big spring in the Tule Narrows.

The ruin impression was fading in the American mind by the time the next artist-explorers penetrated the region. The 1876 Ruffner expedition included Carl J. A. Hunnius, a native of Leipzig, as civilian topographer and official illustrator. Unfortunately, if the extant pen-and-charcoal sketches in his journal are representative, the talent gulf between Hunnius and an illustrator like the incomparable William H. Holmes, who was doing his photo-realistic panoramas of the Colorado Plateau during the same decade, was as wide as the canyons they drew.

Ignore the comparison with Holmes, and Hunnius is still disappointing. He was impressed with the Llano canyons. He regarded the North Fork Draw as "exceedingly beautiful and picturesque," the canyon of McClellan Creek as "magnificent," and the "Grand Canyon" as "sublime" and "the grandest sight" he'd been exposed to in America. But he worried, I think with good reason, that his skills weren't up to the task. Not until 1985 did any of Hunnius's sketches appear in print, and the conclusion could be drawn that there's little reason to violate his obscurity further.

It may be that Ruffner's work is another matter. The young lieutenant, a magna cum laude engineering graduate of West Point, also liked to paint, and he regarded the expedition as an opportunity to try his watercoloring skills on a countryside "so washed and twisted shapen as to marvel the eye with its intricacy and daze it with its brilliancy . . . [these are] wild and enchanting scenes." For his official report Ruffner completed six watercolors, the first colored paintings of Palo Duro Canyon, whose hues elicited laughter and disbelief when he showed

Canyonscape as psychological substitute for Old World architecture and history: Border of El Llano Estacado, *Marcy Expedition lithograph, 1852*

them in eastern Kansas. One of Ruffner's watercolors was of the Tule Narrows, the initial sketch done when he persuaded Billy Dixon to take him into that 700-foot-deep gorge, by no means an easy descent. Ruffner's watercolors are the tantalizing missing file of early Palo Duro paintings. What exist today are his descriptions of them, because the watercolors were not published with his report. Maybe someone in Washington decided that the young lieutenant had let either the heat or his Romantic notions overpower him. I can't prove it, but I have the suspicion that Ruffner's paintings were never published because easterners refused to believe that Palo Duro could be so vividly colored.

ENTREPRENEUR though he was, Charles Goodnight had an eye for beauty in animals and the picturesque in nature. As J. Evetts Haley, surely Goodnight's equal in cold-eyed utilitarianism, has written, Goodnight's life in Palo Duro Canyon actually inspired the old rancher to some of the same observations and sentiments of modern environmentalism. Standing across the Prairie Dog Town Fork of the Red from the location of Goodnight's Old Home Ranch, constructed in 1876 after his cowboys had chased New Mexican *pastores* out of one end of the canyon and thundering herds of bison and frightened black bears out the other, I wonder a little less at that startling claim. No one, not even hard-bitten Charles Goodnight, could watch bright clouds building over these layered cliffs or witness sunset light illuminating terracotta, lavender, orange, and salmon formations in turn without experiencing a few of the feelings that have inspired Americans to want to preserve certain natural spots against progress and human alteration.

So I am thinking this cool July day as Blake Morris and I shoulder our packs, roll up the map, and set out upcanyon. We had gotten a late start from the Claude Crossing this morning and then spent too much time at an immense bankside Chickasaw plum thicket, its branches bent to the sand from the weight of the unharvested fruit. Now we are hurrying, our destination for this night's camp the mouth of Cita Creek, 12 miles (of mostly wading the river) upstream from the car. Twelve miles of cottonwoods and salt-cedar thickets, of coyotes gaping in disbelief, of wild turkeys thumping up and then gliding on set wings across the river, of Mississippi kites by the hundreds circling for insects, of café-au-lait water swirling thickly about our knees. And biting cedar flies to remind us this isn't a stroll in a city park. And a four-inch-long magenta spearpoint reflecting sunlight from the wet sand along the river.

I am not an arrowhead hunter. I have been around archaeologists enough to have absorbed some of their anger and disgust with treasure hunters who destroy invaluable sites looking for artifacts to collect or sell. But I had plucked that spearpoint, a beautifully worked hunting tool of local Tecovas flint, out of the sand on pure reflex and only later reasoned that in that location it could not have been part of a site, instead had probably been years rolling down the river, passing directly from the hand of some Archaic hunter into mine after the modest interval of . . . twenty centuries? Fifty? Seventy?

Evening is upon us by the time the 800-foot canyon walls, which have gradually been pinching inward all day, abruptly swing apart to let us know we are near the mouth of the Cita Canyons, at the extreme south end of Palo Duro Canyon State Park. We are tired and sand covered and the sleeping bags feel awfully good, but before collapsing by the fire we sip coffee and trade notes on this magnificent 60-mile gorge. In two days Blake will leave for Washington to examine National Park Service records for the story behind Palo Duro's ultimate failure to acquire national park status.

Why it didn't is the sort of question that occurs to you if you've spent much time in the wilder parts of this canyon. In a late October, I have walked the 25 miles of lower Palo Duro, from the Claude Crossing to the mouth of Tule Canyon, watching the canyon walls recede until they could only occasionally be glimpsed from the river, lining a course from one windmill to the next because whenever we tried to drink the beautifully transparent river water it spewed reflexively from our mouths like geysers and the gypsum taste lingered for an hour. I have also strolled the broad "hiking" trail through Little Sunday Canyon in the park, most enjoyably in a drenching rainstorm after a peaceful night sleeping on the sandstone bench beneath the Lighthouse spire turned thrilling when lightning began to play about at dawn. The colors of the Sunday canyons are widely regarded by landscape aficionados as more vivid than those of Arizona's Painted Desert, and, in fact, the geology of the two places is identical. But beautiful as the canyons are, there is only one hiking trail, and it suffers from being loved to death by the more than 400,000 yearly park visitors. One fine fall morning, returning from a dawn hike to the Lighthouse and indulging the assumption that I had Little Sunday Canyon to myself, I nearly caught up to a young woman and two children descending the trail in front of me before noticing that the woman was stripped to the waist. It was so natural a response to autumn in Palo Duro that I had done the same more than an hour before. But she knew more about the Lighthouse Trail than I did and shortly slipped her T-shirt back on. Over the next mile, we met more than forty people on the trail. Wilderness—and the freedom it implies—is in short supply in overcrowded Palo Duro Canyon State Park.

The state park was developed with National Park Service help at a time when people like environmental activists Bob Marshall and Aldo Leopold had given up on the park service's commitment to wilderness values. Nothing about the way Palo Duro State Park is set up would indicate that anyone ever anticipated that some visitors would want to hike the back parts of it, and the rangers were surprised when a few hardy souls began in the 1960s to descend North Cita Canyon and then follow the Prairie Dog Town Fork north to the park road, a hike of about 12 miles that requires a lot of brush crashing and only a little trespassing. The state parks people still haven't taken the hint and no system of backcountry trails yet exists in the park.

Maybe it is just as well. Much of the best country in Palo Duro Canyon is outside the park, anyway. In the 1930s and 1940s, when the National Park Service was considering various plans to make Palo Duro either a park or a monument in the national system, even the most conservative idea postulated an area ten times the size of the 14,000-acre state park. Dreamland Falls, which now sits at the bottom of Lake Tanglewood, was to be included, along with a section of the upper canyon heavily used by wintering bald eagles. The erosional badland called the Devil's Kitchen, a pair of large spring waterfall complexes known as the Blue Springs, and a large relict grove of Rocky Mountain junipers were planned for protection—all this above the present park. Below it, historical import necessitated the inclusion of Goodnight's Old Home Ranch site and the beautiful "hole" at the mouth of the Cita Canyons, the location of the famous Battle of Palo Duro Canyon that ended Indian life on the Southern Plains. Evidently it was originally thought that the state park would merely be a nucleus around which a much larger national area would be set aside, because even the Lighthouse, the most famous formation in the canyon, was not a part of the state park and required a separate purchase in the 1960s.

Now most of these places go unseen except by cowboys and stubborn characters such as Blake and me, who are willing to make long walks, wade the river, and pick through trail-less country. It is damned fortunate that at least some of the painters—Georgia O'Keeffe, Frank Reaugh, Alexandre Hogue, and Amy Winton—have liked to hike and camp, or most of us would know only a tiny slice of Palo Duro.

BY THE TURN of the twentieth century the Llanos canyonlands were about to receive the attention of three artists whose work is as close as the Southern Plains comes to having inspired a landscape art tradition. Their works display a diversity of motivations and styles. Under their scrutiny the Southern Plains are as much a human construct, an image of the mind, as is human history—a shifting point illuminated by a prism of distinctive visions. One of those visions represented a southwestern adaptation of a painterly nineteenth century Romantic style much in contrast to the noisiness of the "opera" painters. Another discerned a spiritual home in Palo Duro Canyon and would go on to create a transcendental style that still defies classification. The third went to Taos and absorbed an almost pantheistic infusion before returning to Texas to critically explore modern civilization's relationship to the plains environment. The visions are attached to the names Frank Reaugh, Georgia O'Keeffe, and Alexandre Hogue.

Reaugh is now the relative unknown of the group and even people who have heard of him think of him as "that Dallas artist who painted cows." Two of the three major books on him use "Longhorns" in their titles; Reaugh's somewhat dubious affection for bovines has thus come back to haunt him. Unknown to most is that Frank Reaugh painted between 5,000 and 7,000 tiny pastels, most no larger than 3 × 7 inches, which are almost exclusively landscapes.

Reaugh was a native of Illinois whose parents brought him to Texas in the 1870s. His training included a few months in Europe, where he came under the influence of the early Impressionists and, evidently, that of the Dutch Luminists. Romantic Luminism existed alongside opera painting as nineteenth-century art struggled with ways to express the ability of wild scenery to transport the viewer to a state of religious awareness. What Albert Bierstadt and Frederick Church tried to do with sheer scale, storms, and roiling cataracts, Luminists like Martin Johnson Heade aimed to achieve through scenes of transcendent quiet and repose painted on tiny canvasses, slow horizontal minuets of unruffled water and contemplative mood.

Two tiny Frank Reaugh pastels capture the transcendental repose of the Tule Canyon Basin.
LEFT, Reflected Skirts *and* *RIGHT,* Tulia Cliffs in P.M.
Courtesy Panhandle-Plains Historical Society, Canyon, Texas.

Reaugh is the only major painter of the plains to try this approach and in his best landscapes, few of which have been seen by modern audiences, he succeeds admirably. Beginning in 1889 he made the first of dozens of "sketching tours" to the scenic canyonlands of West Texas. At first he traveled by wagon, later in a Land-Rover type Model T (the "Cicada," he dubbed it), although as late as 1929 and 1931 he was still using wagons to transport groups of Bohemian-looking Dallas art students, plus painting and camping gear, into the rugged canyons around Silverton. Reaugh preferred pastels because with them he could rapidly capture fleeting atmospheric moods, and he concocted a secret formula for mixing colors that were true to this country. He usually painted from nature, using another invention, a combination stool and easel with an umbrella affixed for shade on those summer days when the plains sun cracks open the very ground. Among Reaugh's papers at the museum in Canyon, I've flipped through a surprising number of tiny photographs, mostly of Palo Duro, that he seems to have relied on to jog his memory of the spirit of the place.

With his wild hair and flowing beard he must have looked, and maybe even acted, like some West Texas John Muir with paints, hiking and camping and painting, over a period spanning the 1890s to the 1930s, almost every picturesque scene along the Caprock Escarpment. His favorite haunts seem to have been Palo Duro and Tule canyons. A typical perspective was from a canyon rim overlooking a valley floor devoid of anything human, a completely wild landscape, which Reaugh's impressionistic vision renders as inviting rather than threatening. As in *Lower Palo Duro,* for example, or *Yellow Over Orange.* Sometimes—*The Lighthouse, Fortress Cliff,* and *Tulia Cliffs in P.M.* all illustrate the

point—the perspective is of towering erosional spires or walls from the canyon floor, in the case of the latter painting, the organpipe fins in the Tule Canyon Basin. But judged by the standards of Romantic Luminism, Reaugh's most successful paintings were probably those affecting the tints of the southwestern deserts reflected from the placid shallows of plains rivers. If Reaugh's star as a West Texas romantic landscape artist should ever rise, and it may, it will ride on his ability to open windows into his own soul as he marveled at water and canyon coloring and produced *Reflected Skirts,* or *Orange and Pink Banks.*

Then there was Georgia O'Keeffe.

O'Keeffe had already spent two years in Amarillo, from 1912 to 1914 as a public school teacher, before coming to teach at West Texas State Normal College in Canyon in August 1916. Among her influences prior to moving to Texas was the Russian abstractionist Wassily Kandinsky, author of *Concerning the Spiritual in Art.* Kandinsky's discussion of the way an artist could use certain colors to "cause vibrations in the soul" struck O'Keeffe as an approach particularly suited to the hues and light of the western plains, and in her Texas period she began to perfect her use of color as an emotive force. Eastern artists with whom she associated during the summers soon were telling her that her colors were "impossible"; she responded that they had never seen the country she was painting.

And the country stunned her, even if the people, especially the stuffy faculty who taught but knew nothing about "livingness," didn't. "I'm so glad I'm out here—I can't tell you how much I like it . . . there is something about it that just makes you glad you're living here" she wrote Anita Pollitzer. The "something" was the sky ("Anita you have never seen the SKY—it's wonderful") and nearby Palo Duro Canyon ("Last night I couldn't sleep until after four in the morning—I had been out to the canyon all afternoon . . . wonderful color—I wish I could tell you how big"). It was "a pity to disfigure such wonderful country with people of any kind," she wrote Alfred Stieglitz, but she would ignore them and try to "keep my head up above these little houses and know more of the plains and the big country than the little people."

She did ignore the people as long as she could. She hiked and she painted, in all, nearly fifty pieces from the Southern Plains, although half were blown off her balcony and through the streets of New York City during a windstorm in 1918. Most of the ones that survive are skyscapes. The *Evening Star* series, *Starlight Night, From the Plains I, Light Coming on the Plains,* and *Orange and Red Streak* explored the impact the stunning western skies had on her; they seem to fall just short of pure abstraction only because they are anchored to the earth in the very bottom of the view. Yet, anyone who has watched the plains skies carefully will tell you that O'Keeffe's skyscapes are not abstractions, not exaggerations in any way except perhaps in relative scale.

Fewer of her canyon paintings survive, but the ones that do leave no doubts that her famous "desert" style and her lifelong love affair with the arid, the barren, the eroded were first stirred by the exhiliration she felt in Palo Duro, a canyon she would refer to during her only known return in 1938 as "my spiritual home."

O'Keeffe saw Palo Duro as a magical world of simple yet powerful forms and hot, primary colors. It was also an especially fine place to climb around in, and she found herself going there almost every weekend. "It was colorful," she wrote. "Like a small Grand Canyon . . . wonderful . . . and not like anything I had known before." The prize of her New York exhibit at Stieglitz's studio 291 in 1918, the oil painting she called *Special No. 21,* remains today the most stirring tribute ever made to the blast-furnace effect of the big canyon under certain lighting. In watercolors like *Canyon with Crows* and the little-known *Red Mesa,* she demonstrated her early fascination with lavender southwestern light and its emotional potential in combination with red sandstone and green junipers. "It is absurd how much I love this country," she wrote.

It still shows.

But finally the people couldn't be ignored. Wildest Palo Duro was fenced off, and when a senior art student began to guide her to less accessible places, campus authorities warned him that he wouldn't graduate if he continued to be seen with her. The war was on, and when word spread that she encouraged her students not to buy some ridiculous anti-German postcards on sale locally, she was targeted as unpatriotic. West Texas conservatism routinely banishes the region's most creative people, but this boggles the mind at what could have been. Shortly after, she took a leave of absence from the college and never returned to the faculty.

Years later, the student guide who had shown her places that resulted in timeless national treasures like *Canyon with Crows* had the last word on what O'Keeffe's few months in Canyon had meant. "Did you ever see the rain with Georgia?" he asked. "Did you ever see her watch a great storm? I knew and loved that country well and here, for the first time, was someone who felt the same way about it." It's a feeling that yet shines through in the artistic tribute she left to the Llanos.

WHAT PEOPLE SEE and don't see is interesting. Frank Reaugh had seen the Llanos canyonlands while they were yet pristine and lived to see great changes in the face of the land, but he never addressed them. O'Keeffe had ignored the homesteading that was taking place around her and focused on earth and sky. Alexandre Hogue, who became the force and formulator of the Texas Regionalism movement of the 1930s, could not ignore what the new plains culture was doing to itself and to a landscape he cared about.

As a young man Hogue returned from a calligraphy career in New York to his hometown of Dallas in 1925 to study, at least in part, under Frank Reaugh. There were Taos summers in Hogue's development, too, fine days spent in inspirational company on the west side of the Sangres. But by the late 1920s Hogue had been turned off by the "pretty stagey things around Taos," preferring a Texas cultural scene that was falling under the influence of people like J. Frank Dobie and Henry Nash Smith, who wanted to liberate Texas of its ties to the Old South

Georgia O'Keeffe's Red Mesa *(1917), a little-known watercolor, explored the emotional potential of southwestern light and the colors and forms of the canyonlands.*
Copyright 1987 Estate of Georgia O'Keeffe. Photo by: Malcolm Varon, N.Y.C. © 1987.

Alexandre Hogue's canyonlands vision was pantheistic and environmentalist, as in Red Earth Canyon. *Courtesy Oklahoma State Art Collection, Oklahoma City, Oklahoma.*

and create a new western state culture out of the interaction of Texans with the Texas landscape. More and more, as Hogue and Dallas artists like Jerry Bywaters, William Lester, Harry Carnohan, and Perry Nichols began to create a sort of 1960s-style bioregional movement, Hogue became convinced that the plains-canyonlands country of West Texas would make the perfect landscape symbol of the westernization of Texas. A good writer, Hogue heaped opprobrium on the "mediocre paintings of Texas wildflowers" associated with the art of the Hill Country. But "the rolling plains," he wrote in 1933, "are like still waters, they run deep, but under the surface are extraordinary possibilities for the painter who masters interpretation of them."

In the 1920s it was the tiered canyons, Palo Duro and Tule in particular, which fascinated him most. In 1926 he spent ten days camping and painting in deep Palo Duro, "The Paradise of the Panhandle" he called it in the *Dallas Times-Herald*. "The Canyon wall seems to have had vermillion, yellow ochre and green poured here and there by the hand of a giant and painted in streaks along its face," he marveled. Particularly appealing to Hogue were the relict stands of towering Rocky Mountain junipers that a cowboy who "loved them like they were his own children" showed him. Hogue and others who had seen the canyon convinced Taos artist Victor Higgins, one of the original Los Ochos of the New Mexico School, that Palo Duro was a plains landscape that belonged in Higgins' southwestern portfolio. The result was a marvelous Higgins oil, the perspective from the rim and, appropriately, less than intimate, but the texture of the scene literally translated Hogue's word-picture of Palo Duro into paint.

Before his Taos summers, but especially so afterward, Hogue sensed the truths in naturalistic animism. During his childhood his mother had told him of a "Mother Earth" who lived just beneath the ground. The idea inspired some of the most famous of his acclaimed Dust Bowl, or erosion, paintings. But he began by working the image into "Mother Earth" drawings in the 1920s, showing water erosion uncovering the regenerative sexuality of an earth goddess in a slumping canyonlike scene. This was just the sort of pantheism that was likely to make Hogue critical of the dull roar of factory farming drifting down from atop the Llano and "progress" as it was being defined in Amarillo and Lubbock in those years.

The beginning of his critique of West Texas exploitation came in the 1932 tempera painting called *Red Earth Canyon,* a stylized Tule Canyon encroached upon by power poles that cross the gorge to a farmhouse on the far rim. But it was the Dust Bowl works that brought Hogue national acclaim and a spread in *Life* magazine. Simultaneously, they earned him the status of favorite target of insulted West Texas newspaper editors, who preferred to pretend that there was no such thing as a Dust Bowl and that farming had nothing to do with the sixty dust storms a year the Llanos averaged in the mid-1930s. Hogue never blinked, and in paintings like *End of the Trail, Drouth Stricken Area,* and *Avalanche by Wind* he looked cold-eyed not just at cause and consequence but also at the symbols and technology of the Dust Bowl: plows and barbed-wire fences.

THE OTHER Texas Regionalist painters examined the relationship between American culture and nature on the plains, too, often portraying what a critic labeled "that hard and powerful landscape" with a surrealistic style. But the world was changing. World War II brought an end to the New Deal's WPA art program and with it the official encouragement of regionalism. In 1941 the Llanos got nearly 45 inches of rain and the new irrigation culture augmented that by sucking down the Ogallala Aquifer as the mechanized roar set in once more. The unifying effects of the war and of television were a cloudburst on eroding regionalism and made the nation more homogenous than ever before. The creation of Big Bend and Guadalupe Mountains national parks in Texas soon diverted the attention of artists like Hogue and Jerry Bywaters to the Trans-Pecos, and with the failure of the National Park Service to make 135,000 acres of Palo Duro into a national monument in the 1940s, Hogue's hope that Palo Duro would become the landscape icon of Texas receded from view like a ghostly apparition of the frontier vanishing act. Maybe 1950s' western movies, like John Ford's *The Searchers,* could have arrested the trend if they had been filmed in historical locations. Instead, John Wayne punished his Comanches for the sin of sensuality in a country that bore a remarkable resemblance to Monument Valley, Utah.

DAY IS BREAKING in deepest Palo Duro Canyon. Ke-che-ah-que-hono, or Prairie Dog Town Fork, is reflecting the light of a setting moon and of Jupiter, a morning sky object this summer of 1987. My first sight on waking an hour ago, the star cluster known as the Pleiades, is faint now, and the fire is popping, and while Blake sleeps I am indulging my imagination over a first cup of hot coffee. Watching almost any fire and any sky is worthwhile, but this is a particularly poignant place, especially while running one's fingers over a spearpoint symbolizing 11,000 years of human history, because it was precisely here, just a minute of time ago (about 113 years), that the native continuum ended on the Southern Plains.

Cardinals are whistling now and the cottonwood leaves tell me that the night wind is dying. I step away from the fire and walk in bare feet down to the river, past deer and turkey tracks in the sand. Ankle-deep in the river, looking up through the trees at the moon, I think of a fall camp I once made with the Kiowas over in the Tonkawa Hills. The big Indian camp here 113 years ago would have had that same feel at dawn.

Then, it was late September, farther along in the year, and the moon would have been about the same as the one I'm watching, but in 1874 Comanche, Kiowa, Cheyenne, and Arapaho lodges were scattered through these cottonwood groves, more than three hundred of them extending two miles up Cita Creek. About now the Indians would have been enjoying their last peaceful hours on this sacred ground. Now it is a lonely place. Blake and I are camped less than 150 feet from where Mackenzie's troopers came off the south wall and into the big village. It is paradoxical that I find this place so hauntingly lovely considering what happened here, but I do. It is so hemmed in by high walls and

A. Gormley

A trio of modern visions of Palo Duro Canyon.
TOP LEFT, *Amy Winton's* Palo Duro Summer.
Courtesy Amy Winton.
BOTTOM, *Ben Carleton Mead's* Battle Site at the Junction of Ceta Blanca and Palo Duro Canyon.
Courtesy First National Bank, Amarillo, Texas.
RIGHT, *Daryl Howard's* Palo Duro . . . Calling the Sun I, II, and III.
Courtesy Daryl Howard, Inc., Austin, Texas.

nestled in the soothing rustle of cottonwoods that it must have seemed to the Indians that they had entered the very womb of the Earth Mother here and could not possibly be harmed. In fact, Maman-ti, the Kiowa leader whose band had put up its lodges about where we are camped, had brought them here because of a vision that in Palo Duro's embrace the Kiowas would never be found by the whites.

The crows are announcing it: half an hour of twilight left.

I spend sunrise and the first two hours of full light exploring Cita Creek and the hole created in Palo Duro by the merger of the two streams. I have plenty of time; we have the rest of the day to make the state park campgrounds, six miles upcanyon. The Cita is shallow but like the main stream is clear and runs over rocks in this section, and the water is good to the taste. Stepping over turkey tracks, I move along the bends where the Cheyennes had been camped, under Iron Shirt, and the descendants of the kestrels that circled Cheyenne lodges peer down at my intrusion. On high ground northwest of the confluence I sit atop a boulder and look across the tree-lined creek where O-ha-ma-tai's Comanche lodges were being sacked and burned by this hour of the morning. Now I realize, maybe more fully than I ever have, my own strangeness in West Texas, for the image that resonates for me is not that of American progress, plunging down the canyon wall in the form of blue-coated troopers, but of the last peaceful hours of the great village—dogs barking, horses nickering, woodsmoke rising, the muffled laughter of the unknowing doomed.

I AM NOT ALONE in that sense of identification, of course. Images from the frontier and its end are weighty icons for modern Southern Plains culture, and Remington-Russell-type scenes still emanate in an unending stream from the hands of local artists (Harold Bugbee is considered the best of them by aficionados of cowboy art). But it's the work of Don Ray, a Channing, Texas, landscapist who knows history and likes to paint early plains explorers, and that of the late Ben Carleton Mead, a muralist and illustrator of J. Frank Dobie's books, whom I once talked to at length without knowing who he was, that can stop me in my tracks. Ray's large oils, such as *Last Light, Palo Duro* (1985), or *Spring in the Canyon* (1986), seem nineteenth century representational at first glance. Closer experience with his paintings discloses a sensitivity to light (I once heard him describe "those quiet autumn afternoons" on the Llanos "when the light is so strong you can hardly look at it") and the subtleties of nature that reveal a cowboy upbringing tied to a Romantic's soul. As for Mead, although he rarely painted subjects that have attracted modern Native Americans, until Indian artists rediscover the Llanos canyonlands, his is the most sympathetic vision of Indians and their time. Mead's best Palo Duro oil freezes the last moments of the free Plains Indian life in a glowing panorama of the Kiowa, Cheyenne, and Comanche village at dawn. To paint it, he visited the rimrock overlook of the spot on the date of the battle for three years before doing the final painting on September 28, 1974, the centennial of the Indian armageddon. Mead's most evocative historical painting, however, happens to be the most Indian-like work he ever did. He called it *Comanche Horses of Tule Canyon* (1978), and it fixes in the mind a haunting, surrealistic image of the slaughter of the Indian pony herd in upper Tule, with riderless ghost ponies floating over the canyon and into plains mythology under a pale, chill Comanche moon.

OF AN EARLY SEPTEMBER, Amy Winton and I hike up Mulberry Canyon past bankside fall wild flowers, pale trumpets and fiery cardinal flowers and here and there clumps of blue penstemons that aren't supposed to be flowering here now but obviously are, looking for a brightly striped wall I think she will want to paint but which I am having trouble relocating. A transplanted New Englander, Amy is a study in regional conversion. For seven years, she tells me as we work our way up the creek, she resisted everything about West Texas. But Palo Duro ate away at the distance she maintained and finally seduced her. Along with her friend Daryl Howard of Austin, she is today the best known of the modern painters of Palo Duro and has a rimrock studio there.

Now, like her hero Frank Reaugh, she is broadening her scope, painting Tule and the canyons of the Little Red and the shallower drainages north of Palo Duro, of which Mulberry Canyon is the most picturesque. Mulberry is, in fact, Amy's kind of canyon: intimate, not especially grand or awesome, the sort of place many people would overlook. But enter into it and an unsuspected world opens up where rare flowers bloom and the dwarf shin oaks grow into stately 30-foot monster trees. For good reason the Comancheros called this canyon Palo Grande; it still has the most widespread forest of Rocky Mountain junipers in the Llanos canyonlands.

A mile from where we left the Jeep I still can't find the particular clay wall I want her to see, but there are others and we photograph them and talk. These are not the kind of scenes, I suggest, that would appeal to the grander vision of a Wilson Hurley or a Daryl Howard. Both of these modern landscapists belong to the Canyons School, or Rocky Mountain School, of the greater American West. Hurley's marvelous *From the West Rim (Palo Duro Cañon),* demonstrates his affinity for panoramic views and the kind of scale that appealed to Bierstadt and Moran. Daryl Howard, in contrast, seeks the power of simple forms, which she renders in a complex wood-block style synthesized from her Japanese and Hopi teachers. Daryl's involvement with the Llanos canyonlands dates to 1981, when she first painted Palo Duro and Tule and decided to return at five-year intervals. *Canyon Morning* (1981) and *Palo Duro . . . Calling the Sun I, II, and III* (1986) are wonderful contemporary landscapes that recall O'Keeffe in their minimalism and sky emphasis. Gaining access to the other, privately owned canyons of the Llanos is a particular problem for modern painters, but in 1985 Daryl did manage to get into Yellow House Canyon. The resulting works, *Plains Secret I and II* and *Secret Journey,* establish a bridge in both landform similarities and painterly style with the Four Corners country and the paintings of Ed Mell.

Amy's vision is different: "I paint these canyons because I really want people to know about them, to pay attention to things they would walk past without noticing otherwise. Sometimes I think I do it because I want to test the feeling in myself, to affirm the reality of what a place says to me by painting it and then showing it to someone to see if *they* feel it—if the light, the colors, the smells that make my heart sing when I'm there—is something a New Yorker or a European will react to similarly. If so, then I'm affirmed. I'm not a maniac wacko or a hopeless Romantic . . . or at least not the *only* one."

Of course she is not, but of course she is a Romantic. As is Daryl Howard, as is Don Ray . . . and as am I, appreciator of their soul visions of the wild scenes these canyons yet preserve. As a guide, however, I am a bust. Two miles upstream I am forced to conclude that the terracotta wall mottled with green and yellow and magenta that I want to show her must be downstream instead. Or else it must be in some canyon of my imagination.

Chapter 7

Dust-Blown Dreams and the Canadian River Gorge

DISAPPEARING into a thin, red horizon line off to the west is a squashed, third-quarter moon, getting as broad and bloated as it sets as a country-boy beer gut ballooning over sagging Wranglers. I am just north of Clovis, the Allsups Convenience Store Capital of the World (with good reason—Lonnie and Barbara Allsup are native Clovisians), whooshing along the New Mexico Llanos the way we late-twentieth-century folk do. It is January in the Siberia of the Southwest, 18°F at 5:00 A.M. Rocky Mountain Time. I am running an errand, fetching a load of piñon for my wood stove, an excuse, really, to slip over into New Mexico and poke around the Canadian Gorge for a few hours.

The plains are almost wholly unfit for cultivation, and, of course, uninhabitable by a people depending upon agriculture for their subsistence.

—Explorer Stephen Long, 1820

But dost Thou know that for the sake of that earthly bread the spirit of the earth will rise up against Thee and will strive with Thee and overcome Thee?"

—Fyodor Dostoyevsky, *The Brothers Karamazov*

So long, it's been good to know you.

—Woody Guthrie, 1938

The gorge. In New Mexico that name conjures the Rio Grande, trout fishing, Wild and Scenic River float trips, whitewater, and the stretch known as the Box. Not many would connect it with the inches-deep Canadian River slogging its way across the plains east of the Sangres. Once, at a highbrow backpacking store in Santa Fe, I looked through their topographical map file for one of the Canadian Gorge. Zilch. The Land of Exploited Enchantment has too much else to offer to concern itself with a hard-to-get-to plains canyon. So the Canadian Gorge is one more of those out-of-the-way places in the West that one never hears about. But best you find your own. This one's taken already.

Like most of the plains canyons over in Texas, the Canadian Gorge and its jangled maze of side canyons are lost beneath the horizon in a remote country with few roads. Most plains people know of the Canadian River, whose two forks drain across Texas and much of Oklahoma. But hardly anyone knows that the main (south) river cuts a deep canyon into the plains, or that the river itself, Río de la Cañada Colorado, was named that by the New Mexicans for this *gran cañada.* The first time I heard anything about it was in the summer of 1979, when I was doing the bicycle trip I've elsewhere mentioned across the Llanos

OPPOSITE PAGE:
On the rim of the Gorge, the Canadian River far below

to Colorado. After I wheezed up the thousand-foot Canadian Escarpment east of Mosquero, Ricardo's Bar in Roy seemed like a good idea. Friendly folks. If I was climbed out today, someone offered with a wide grin, best not ride west out of Roy. BIG canyon. I took the advice but remembered the characterization. A couple of years later, when I finally drove the Roy–to–Wagon Mound highway, I was impressed, too. The Canadian Gorge is, simply, the biggest and deepest canyon on the American Great Plains.

In predawn light this frosty morning I drop off the Llano Estacado past Caprock Amphitheatre State Park, where cowboy musicals are performed in the summer. Past San Jon, the basic physiography of the region sharpens. I am in the breaks between the two Llanos, their escarpments standing 30 miles apart and both visible, as is the familiar layer cake of Tucumcari Mesa down on the southern horizon. (Tucumcari is possibly a Comanche term meaning "to lie in wait for someone to approach," although some sources believe the name comes from the Apache word for "breast"—just tilt your head to the horizontal.) To the west the white sandstone columns at the base of the Canadian Escarpment glow magenta in the winter sunrise. The truck breaks out of the blue light, its shadow rippling along the base of the scarp. Then comes the long pull up, a glowing vastness in the mirror and, all of a sudden, the mountains, snowcapped and one-dimensional as cardboard cutouts, 60 miles away.

And, astonishingly, there is the gorge, too, hanging high in the air, its west wall clearly visible although the details are wavering as if seen through water. It's the winter mirage, the phenomenon of cold, still mornings on the plains, bending light downward so you can look into the canyons, and the best expression of it I've ever seen.

It's not an apparition; the Canadian Gorge is real. But apparitions from half a century ago still loom atop the Mosquero Flat. Ramshackle farmhouses, crumbling to dust, hove in the distance. Yellow grasses carpet the plain, but beneath them, like corduroy, are the unmistakable lines of what were once furrows throwing shadows in the low-angle sunlight. This country was once part of the Llanos agricultural boom.

A raven flaps across the plain. With the windows up against the chill I can't hear what it has to say, but it's not hard to guess.

Four hours later I am bumping along a dirt road, climbing out of the mouth of the gorge through bajada slopes of black lava blocks and lime green prickly pear, past a spray-painted "Far Out" on a big boulder, leaving Sabinoso and Largo Canyon and the swift, turquoise blue river behind.

It had been intended as a quick side trip to check out a theory. Although I have hiked the upper half of the Canadian Gorge, I'd not seen the mouth of the canyon where the main gorge is joined by Mora and Largo canyons, a coiling junction 1,200 feet below the rimrocks. For years I have been convincing myself that one can float through the greater part of the gorge's 45-mile length at some water level below outright flood; this trip was supposed to be a float trip homework. But prowling around Sabinoso, a loose little Hispanic farming and ranching community strung out for three or four miles along the river, I had become intrigued instead by the images of people who had stuck, who had adapted to the harsh, isolated life of the gorge. They had been here, had made this part of the West their living space, for more than a century. Yet the farmers up top and those farther upcanyon hadn't lasted four decades. What was the explanation for that?

IT IS THE WEST'S grandest illusion, foisted on a population that wants to believe it by historians who made their names saying it, that the pockets of American culture that have sprouted like mushrooms across the American West over the last century have successfully adapted to the environment of this country.

One of the best-known books about the American West is Walter Prescott Webb's *The Great Plains: A Study in Institutions and Environment,* which was first published in 1931. Three generations of plains people have now grown up reading it and accepting Webb's repeated premise that their culture has been shaped by the Great Plains environment in far-reaching ways, even to the point (as a recent president of the Southern Plains' major university, now a national political figure, would have it) that "like the plains themselves" the people are especially "open," since there is "nothing to hide behind." Meanwhile, plains people have precisely the same culture as people have in the rest of the country. They eat the same cereals for breakfast, work in the same service-related jobs, watch the same sitcoms and sports events on the same networks as everyone else, produce agricultural products for an international market economy in the same way that corn farmers in Iowa or cattle ranchers in Louisiana do. If, out the windows of their farmhouses or Fords and Chevies and Buicks and Cadillacs, they see barbed wire and windmills and center-pivot irrigation systems, they remember Webb and conclude that, if such symbols are still in place, then the adaptation must have been a success and must yet be going on. That is, if they think of such things at all. (Webb, naturally, did continue to think of such things. Twenty-five years after *The Great Plains* was published he wrote an essay, "The American West: Perpetual Mirage," in which he set forth revised thinking: that the heart of the West was desert and the regions and towns that ignored that were destined for abbreviated careers.)

The idea that we have adapted to the western environment is a palpable fiction, scarcely true anywhere in the West unless it be among some of the southwestern Indians or the New Mexican Hispanic villagers. And it is a fiction nowhere more transparent than on the Southern Plains. Nowhere else in the West has nature presented a better example of how ill-adapted American culture is to fragile environments than in the dust bowls that periodically drive the Southern Plains to the brink of ecological and economic collapse. And, if you want an example from the historical record, one that portends the probable future for the rest of the Llanos, try Harding County, New Mexico, one of the counties (Mora is the other) lying athwart the Canadian Gorge.

Fed by Rocky Mountain snowmelt and springs seeping from the Jurassic and Triassic formations, the Canadian bounces swiftly over a rocky bed.

THIS IS BELL RANCH COUNTRY, all these eroded mesas and breaks from the mouth of the gorge down the Canadian River nearly to Tucumcari, almost 50 miles away. Founded by Wilson Waddingham in 1870, the Bell Ranch is still one of the largest working ranches on the Southern Plains.

At its northern periphery I bounce onto the highway and turn back toward Mosquero, past La Cinta Canyon with the pretty little Triassic red mesa standing in its portals like a sentinel. Once, I prowled around in La Cinta Canyon and spent an hour talking to Ray Hartley, who owns a ranch there and who has spent his life, except for a few years getting a college degree up in Fort Collins, Colorado, living in the canyon. I asked him about Sabinoso. "Those Mexicans have been in the deep canyons around the mouth of the Mora River forever," he said. "Floods and isolation and the Dust Bowl run everybody else out. But they won't never leave."

Atop the Llano again I turn north toward Roy. Along with Bill, Wyoming, I think Roy, New Mexico, is one of my favorite "towns" on the Great Plains. Roy was named after two brothers from Canada, Frank and Eugene Roy, who arrived in 1901, just after the El Paso and South-

Mouth of La Cinta Canyon

west railroad, which was supposed to haul coal down from Colorado, was surveyed across the grasslands east of the gorge. Sixty-five years earlier, the government in Mexico City had granted all this country to a consortium of individuals who were supposed to settle it. The Mora and Las Vegas grants (1835) and the giant (1.7 million acres) Beaubien and Miranda (or Maxwell) Grant did bring folks to the area. But the New Mexicans were practical settlers. They founded most of their communities, based on small irrigated farms and flocks of sheep, in the gorge and the breaks east of the Canadian Escarpment. Sabinoso, Mosquero, Armento, Gallegos, and others were the result.

When the railroad arrived, the small New Mexican settlements were still alive, but much of the Llanos grasslands surrounding the gorge had meanwhile fallen into the hands of a Portsmouth, England, cattle company or had become public domain when the United States pried the Southwest loose from Mexico. But the Roy brothers had grandiose plans. While Eugene supervised their ranching operation, Frank became the first postmaster, banker and city treasurer, store owner, saloon owner and bartender, and mayor of the town they naturally named after themselves. After the railroad arrived in 1902, Mosquero (it means "swarm of mosquitos") was relocated from the canyon to the rail line south of Roy. Mills, a former frontier stagecoach way station named for the Mills family, whose irrigated ranch had been located in the gorge in the 1870s, was likewise moved atop the Llano north of Roy. These were typical frontier booster decisions: water was important, but, hell, a railroad spelled bonanza, U.S. style.

I pull up to the post office on the southwest side of the little town square. Roy, of course, is dying. There are a couple of groceries, a pharmacy, a café, Ricardo's Bar. Most of the other buildings are empty now, in a town that sixty years ago was the hub of buzzing economic development. Bob Wills, of western swing fame, used to barber here between road trips with the Texas Playboys. But the halcyon days when Roy and Mosquero battled for the county seat of newly created Harding County (it was 1923; Mosquero won, but Roy got the high school) seem as long gone as the buffalo herds. Roy and Harding County have readjusted their visions considerably.

I give my couple of pieces of mail to the postperson and stroll over to the Roy City Market. The mounted heads of aoudads and African ibex, another exotic big-game animal that has been introduced into the gorge, line the walls above rows and rows of Vienna sausages and every flavor and combination of sardines imaginable. It's a sardine Baskin Robbins. I settle for a Coke and walk out into the bright winter sunlight. Beyond the edge of town, beyond the edge of weeds and fences plastered with blown paper, brilliant yellow grasslands stretch away in every direction. The gorge is visible to the west, a dark green slash running north and south through the yellow. Beyond another strip of yellow are the black-looking foothills and then the white tops of the Sangres, and north of them, up in Colorado, are towering snowcapped peaks just now starting to gather clouds.

I start the truck and head north out of Roy and try to imagine the scene here half a century ago. It's not hard. It's still going on over on the Llano Estacado.

MOISTURE determines, it has been said. And those mountains standing high on the western horizon determine the moisture on the plains sweeping away from them. They've decreed that the annual rainfall on the Llano around Roy is only about 15 inches, even less than the 18 or so on the Llano Estacado, since the rain-shadow effect is more pronounced here.

Initially, it was commonplace—at least since Stephen Long came down the Canadian in 1820 and waffled not the slightest about his impression—to conceptualize the Southern Plains as a desert. By ecological definition it is not, at present, but it stands so close that in the years, or decades or centuries, when the bright summer clouds do not build rain-bearing thunderheads, the lack of moisture tips the balance in a decidedly deserty direction.

This is something the New Mexicans evidently figured out for themselves. Aside from a few playa lakes, there was no water up on the Llano. The water was in the gorge, flowing as melted snow down the narrow canyon, leaking out of springs in the side gorges from what is now called the Canadian River Underground Water Basin, a narrow little strip of Ogallala Aquifer that unfortunately doesn't store much water.

But the railroad was in place and boosters were doing what boosters in the West have always done, hoping to drown doubt in a din of chest thumping. New Mexico Territory's Bureau of Immigration encouraged farmers from the East to come and participate in "breaking out" (plains terminology; its opposite is "hairing over") the grasslands in "this princely domain." The Enlarged Homestead Act of 1909 and the promise that wheat could easily be dryland farmed contributed to luring unsuspecting sodbusters to the New Mexico Llano, just as the breakup of the giant ranches, the XIT in particular, did on the Texas side. On the Southern Plains the farming frontier did not end in 1890. It extended well into the 1920s, a twentieth-century frontier confronting a landscape that fifty years before had been regarded as one of the only uninhabitable parts of the West.

So they came, poor pioneers from Texas and the Midwest, following the American dream. Fabiola Cabeza de Baca, whose father was one of the New Mexicans the pioneers had come to supplant, described them in her memoir as arriving in wagons from the railheads, "kindly, simple folks," although one did mightily insult her father, a fair-haired, blue-eyed Spaniard of Galician descent, with the comment, "I thought you were a white man when I saw you." All across the grasslands of the Llanos they settled on their rectangular squares of land, ripped away the grass cover so that gasoline-powered tractors and one-way disk plows could get at the soil below, built homes and raised children and dedicated their lives to making the Llanos yield to agricapitalism. The Big Breakout, forget the Romantic Jeffersonian notions of farmers, made war on nature. Modern farming, particularly plains monoculture farming, with its ammonia phosphate fertilizers, its mined water, its poison chemical pest control, its genetically inbred crops, places farmers in a nonreligious relationship with the land. Their profession gets them about as close to the ancient Earth Goddess religious feeling of early agriculture as today's sport hunters are likely to get to the animal mythologies of the primal hunters. But who can blame farmers? The lesson is held before them again and again: take advantage of technology and go into "business" or you don't get rich. You might not even survive.

Nature's lesson was simpler and, if anything, more cold blooded.

THE PATTERNS of plains weather over geologic time demonstrate regular pulsations spanning long evolutionary cycles, products of the grand forces of the solar system, like sunspot storms and oscillations in the earth's orbit, and of changes in the ocean currents in response to the shifting of the continental plates. Cores taken from the sea floor indicate that over the last million years major warm-dry and wet-cold episodes have supplanted one another at roughly 550,000-year intervals. Within the larger swings are smaller pulses that still powerfully affect biological life and within those yet others governing narrower and narrower slices of time. Pollen analysis and dendrochronology have discovered significant oscillations on a 2,000-to-3,000-year pattern, while our brief meteorological records show weak pulsations at intervals of 40 years (the droughts of the 1890s and 1930s) and 20 years (those of the 1950s and 1970s). Climatologists argue that we are presently in a slow drift in the warmer-dryer direction of one of the large swings. One prediction for the evolution of this stage is droughts spaced gradually closer in time and lingering longer until a 2,000-year peak is approached. What impact the Greenhouse Effect might have toward accelerating this cycle is something no one knows beyond the prediction that its effects will become increasingly noticeable over the next 30 years. But an annual temperature shift of only about 3–4° F would advance Chihuahuan Desert species 200 miles into the Southern Plains. The increased evaporation and solar convection will likely turn much of the southern High Plains into an American version of the Sahara.

Why then, in his prize-winning history *Dust Bowl: The Southern Plains in the 1930s,* would Donald Worster call the Dust Bowl one of the three greatest human-induced ecological collapses in civilization's history? Because, while nature served up an ordinary 40-year drought like the kind that had been disrupting Native American cultures in the Southwest for thousands of years, American culture had made the plains ecologically vulnerable overnight. The drought started on the Northern Plains in 1932, reached the Southern Plains by 1934, and seemed to settle in around the point where Texas, New Mexico, Oklahoma, Kansas, and Colorado join. From 1935 until 1938, instead of getting 15 to 18 inches of rain a year, the Llanos averaged less than 8.

Even where the vegetation is intact, such droughts burn up the plains. Botanical records from native mixed prairies during the Dust Bowl and the drought of the 1950s show far-reaching decreases in diversity. Vegetation survived better in the canyons and breaks, but even here the tall grasses gave way to shorter ones that were then also burned up by the scorching sun. For vast stretches, the only green was provided by Russian thistle, white horse nettle, sunflowers, and mesquite, which seemed to thrive and spread. Full recovery of mixed grass prairies on the Southern Plains takes between 20 and 40 years and four succession stages from annual weeds to a mature mix of grasses, forbs, and shrubs. The impact of regular droughts is a prime reason why "climax" is almost a meaningless botanical term on the Great Plains.

What made the drought of the 1930s into the Dust Bowl was the presence of farmers in a country they shouldn't have been in, doing what was normal for farming but shouldn't have been done in that country. With the grass and herbs and their extensive root systems gone, the Llanos simply blew away. In several places all the topsoil blew off down to the caliche rock. Rather than two or three dust storms a year, in the middle thirties there were sixty or seventy. The stories have become a genuine part of American mythology; in fact, the Dust Bowl is the great historical experience of the Southern Plains, producing tragedy and suffering, also high literature and art and a lasting folk music.

WITH THE VOLCANIC CONE of Capulin Mountain against the northeast sky I turn west toward Springer, past one of my overnight camps when I biked through in 1979, one shared with a Florida lawyer hitchhiking to a tribal gathering of old Sixties friends in Durango. Pronghorn glide across the short verdure, jackrabbits lope across the highway, and here is the Canadian River, shallow braids of water in a sandy channel. This is upstream from the beginning of the gorge, very near the crossing of the Cimarron Cutoff of the Santa Fe Trail, which looped up this way to avoid having to cross the gorge itself, but still 60 miles from the sources of the river up in Colorado. An amazing degree of geographic confusion prevailed about the Canadian as recently as 50 years ago. Settlers during the Big Breakout assumed that they were on the upper Red River of Texas. The Resettlement Administration disabused them of this notion in the late 1930s, but for another decade this was known as the Canadian Red River.

I drive on through Springer and follow I-25 for a short distance before turning off for Cimarron, a tiny burg located where the Cimarron River leaves its mountain canyon, carved down from the high mountain park Texas developers have made into a skiing and boating resort. Despite its far-fetched nineteenth-century claims, Texas never managed to get possession of any of New Mexico except the Llano Estacado and the Trans-Pecos. Sagebrush rebels, who campaign against federally owned land, might take heed from the Texas lesson, for Texas offers an example of what would happen to the rest of the West if public lands were turned over to the states. With almost all of their own scenic lands privately owned, Texans have come to regard New Mexico as a Texas playground. The resulting "dollar imperialism" creates an almost Third World–like, love-hate attitude toward Texas on the part of the native New Mexicans. Introduce yourself as a Texan in Taos or Santa Fe and watch the corners of their mouths.

I make my firewood deal in Cimarron. Reeking of barely seasoned piñon, wallowing on flattened suspension like a drunk, the old Toyota and I go careening back through bright winter sunshine. At Springer I stop at the Stockman's Café, eat a trout covered with green chilis and piñon nuts, swill down cups of steaming coffee. Back on the highway to Roy, bald eagles perch like kingbirds on power poles along the highway.

A few miles from Roy I ease the pickup onto a dirt road and like a ship on an ocean sway along for 10 miles into the approaching sunset. A few minutes before the sun hits the top of the Sangres I pull into a scattering of picturesque piñons and stop. A brown National Forest sign—this is the Kiowa National Grasslands—tells me that the next two miles are exceedingly steep and rough and aren't recommended for passenger vehicles unless the passengers are in search of that dwindling point of light at the end of consciousness. I've driven this road in all kinds of weather, and they're not kidding.

I walk down the road, out of the last gathering sunlight and into the shadow cast by the far wall of the Canadian Gorge. Behind me, through tall ponderosa pines, a waxing moon that is now two days from full hangs over the tawny National Grasslands, with a band of blue nightfall already rimming the horizon.

DROUGHT and agriculture combined to turn enormous stretches of the Llanos into sand dunes in the 1930s. Animals died of silicosis, caused by dust inhalation, and whole counties simply blew away. Outmigration from the Southern Plains, much of it along famed Route 66, approached 20 percent; the plains lost nearly a million people between 1930 and 1940. Franklin Roosevelt's agencies did what they could. Shelterbelts were built along the 100th meridian to contain the desertification. Most of the farmers on the plains were on relief. Roosevelt called for various ideas to resolve the crisis, got suggestions ranging from concreting over the plains to dumping used cars and junk across the region. Ecologists and land-use planners argued that far too much of the plains had lost its grass cover in the move to factory farming, an analysis seconded by the 1937 presidential report *The Future of the Great Plains*. In that year the Bankhead-Jones Act provided funds for the reacquisition of homesteader lands on the Great Plains, and the Resettlement Administration was set up to relocate Dust Bowl families. Homesteading in the West was ended for all time; the remaining lands eventually were administered by the Bureau of Land Management. But two human developments, the massive spending of the Soil Conservation Service and the adaptation of ordinary auto engines to produce a reliable pump for plains irrigation, combined with a natural one to create a regional illusion that technology had conquered the Dust Bowl.

Nature's contribution was the swing toward wetter conditions. The 41 inches of rain in 1940 really broke the Dust Bowl.

THE CANADIAN GORGE is a magnificent maze of remote canyonland country occupying an area of nearly 700 square miles. Scores of side canyons drain into it, almost all of them with names dating to the nineteenth-century period of New Mexican settlement: Mora and Largo canyons are the largest of them, but there are in addition Cañon Biscante, Arroyo Piedra Lumbre, Cañon Colorado, Cañon Vercere, Cañons Mesteño and Mesteñito, Cañon Emplazado, Cañon Blanco, Cañon Capulin, Cañon Armenta, Cañon Encierro, Cañon dos Nieves Gachupín, Cañon Juan Maes, Cañons Osa and Osito, Cañon Hondo, Cañon Yegua, Cabra Cañon. And Mills, Beaver, Whitman, and Davis canyons. West and east along the Canadian Escarpment the mesas and canyons stand on the land as if stamped there by a giant cookie cutter, a landscape of juniper-speckled truncations like Mesa Herfana and Cerro Corazón and rocky reentrant canyons like La Cinta. It's the Caprock Escarpment on the grander scale of the farther west, with the Capulin and Turkey mountain cones and the sheer gray block of Hermit's Peak to rest the eye on when you top the rimrock. These are the names that organize the erosion jumble that is the Canadian Gorge country.

Apparitions out of the past stand in plain view as you hike the Canadian Gorge. The impression is of a great ghost canyon, an empty house where the echoes of human voices have barely fallen silent. Pottery shards and flint scatters lie beneath layers of dust in the overhang caves. Lines of piled rock mark ancient Hispanic sheep-range boundaries. There are overgrown cemeteries and the crumbling plaza of the old abandoned town of Armenta. Bois d'arc trees scatter their pebbly fruit across ground once occupied by southern pioneers. And rock-house ruins are everywhere—of the old Mills Hotel, which once served the stage line; of cowboy line camps; and, more sadly, of failed Resettlement Administration homesteads.

The late-twentieth-century gorge is part wilderness experience, part history lesson. Bill Brown and I hiked it in late September 1986, the upper half that is now in the Kiowa National Grasslands. We left a pickup down at the Roy–to–Wagon Mound highway, the only road that bisects the 45-mile length of the gorge. Katie Dowdy dropped us off at Mills Canyon campground, near the head of the gorge, on her way to doing a rock-art reconnaissance of Eagletail Mesa. If things turned out right, we planned to rendezvous with her at Ricardo's Bar in Roy the next evening.

Excerpts from my journal:

"September 19, 1986: After almost a decade of promising myself this walk, Bill and I sit satisfied beside a popping juniper fire in the gorge of the Canadian River. We set out after lunch today on a 2-day excursion, about all it will take, even allowing for some side-canyon exploration, to hike the 18–20 miles of the upper canyon. We did a leisurely 4 or 5 miles through 'Mills Canyon' this afternoon. It is beautiful, a broad floor carpeted with buffalo grass and miles of smooth sandstone walls striped with the dripping paint stains of desert varnish, mineral residues left when moisture streaks evaporate. According to Katie, a method has recently been worked out for dating the layers, a boon to rock-art research. Unfortunately, salt-cedar thickets line the river here about as thickly as they do anywhere in the Southwest, a boon to the horrendous mosquito hordes that descended on us at dusk. *Tamarix chinensis* tends to thrive especially well upstream of reservoirs, where slowed water flow creates mud flats. And Conchas Reservoir is only about 75 miles down the Canadian.

"We noticed as we walked this afternoon that autumn is touching the gorge, more distinctly nearer the rimrocks, at an elevation of 6,000 ft., than on the warmer canyon floor, 800 ft. lower. But the scrubby wavy-leaf oaks that cover the upper slopes are already tinged with purple and red, big rabbit-brush clumps are a dazzling fall yellow, and even the cottonwood galleries along the river are close to turning. The survivability of plains cottonwoods amazes me. They have evolved chemical defenses against grasshoppers, who will strip every other green thing in a country but pass up cottonwood leaves; they combat aphids by making certain leaf clumps so aphid-attractive that the insects ignore the rest of the tree. All this must be a product of cottonwoods' rapid generational turnover.

"Present camp is on a grassy shelf on the E bank, 20 feet above the river, which splashes along pleasantly as white noise. We're evidently about 1/2 mile below the mouth of Cañon Mesteño and near a cave overhang that we'll explore at first light. I write with the red punctuation point of Mars in the southern sky, Jupiter just over the canyon wall behind me, and a waning near-full moon about to slide into view. Planes, bound for L.A. or New York, sweep over but they're too high to hear above the sound of the river. No bawling stock, not even a coyote. This is an extraordinarily silent canyon. It got to the explorer James W. Abert, who somewhere in this very stretch in 1845 found the scenery lovely—'a smiling valley'—but was disturbed that the great vastness so utterly ignored him. He fired a shot into the gorge just to break its composure. Now, just as I scribble this, a coyote babbles eerily off to the SW; another falsetto-voiced one joins in. It must be a moon serenade. They're up the canyon wall, in the chill blue glow of the moonrise.

"It is all wildly beautiful, at times (when the fire dies and the mosquitos swarm) pretty uncomfortable. But Bill, curled in the fetal position in his bag, is happy. His graduate thesis is on sequential cultures along the Canadian River, and this is his second 'field trip' along it; he's already hiked a long stretch of it in the Texas Panhandle.

"I confess to some partiality to the wonderful but unsung places of the world: canyons like this, unknown, unvisited, with the bark still on, a place you can make your own by dint of experiential immersion. My place. If I write about it, it's not to sing it as a place for everyone to know. We all have to find our own places.

"September 20: Too conscious of the evolving night to sleep well, I crawl out of the bag 2 or 3 times to stoke the fire, sit sleepily beside it as the twigs catch. Once, adjusting my bedroll, I notice in the moonlight a big furry plains tarantula, near cereal-bowl size, waltzing across my chest. Probably a male. We'd noticed big autumn tarantula migrations (usually all male) crossing the roads yesterday, much like the kind, Scott Momaday recalls in *Way to Rainy Mountain,* that in September of 1874 had frightened the Kiowas into joining the doomed Palo Duro village. This one pauses a few inches from my face, scrutinizes me with apparent horror in all 16 compound eyes before scurrying off. Tarantulas are actually pretty passive, with a bite not much more toxic than a wasp sting. Nothing like the poison in the mesquite thorn I jabbed into a finger joint yesterday. A quick application of tequila into the wound is the only antidote I've found to mesquite toxins, and we have no tequila. The joint is so stiff this A.M. I can barely move it.

"Around 1 or 2 A.M. by the Dipper, the wind came in high from downcanyon but by early dawn it has died down. When Sirius rises like a Christmas sparkler over the east rim I blow the fire to life and put on a pot of water. It's warm this morning and the irascible little mosquitos are delighted. They and the coffee smell finally wake Bill, about the time the west wall turns red-gold. In the dawn quiet the sound of rapids downriver fills the aural space. A little plains black-headed snake, no larger than a pencil, crawls across my foot in hot pursuit of a centipede. For two hours after sunup our camp stays in cool blue shadow. A good place to be, this.

"While Bill works on his journal I explore the overhang cave in the wall behind us. It is a large cave, 12 ft. high, 25–30 ft. across at its mouth, maybe 15 ft. deep, and as I approach I see on its ceiling characteristic smoke smudges. The next thing I see is a discarded dig screen. The cave has already been seined by someone. Although the Comanches regarded the gorge as the western boundary of their range,

OPPOSITE PAGE, TOP: Bill Brown contemplating our 20-mile descent of the upper Canadian River Gorge; BOTTOM, A canyon capable of evoking one of the simplest of human urges: to know what lies beyond the next bend.

this was probably an Archaic site. The Canadian Gorge has always looked like prime Anasazi country to me, but I am unaware of any Anasazi sites here.

"The sun lifts higher and soon the riverbed is bright with the reflections of wet rock. A mist rises from the river. The gorge is already starting to trap solar radiation and is heating up, and we decide to move out now and dry the dew from our bags later in the day.

"So it's down the canyon, indulging that simplest of lusts—to know what's there. What does it look like around that bend we've been gazing on since yesterday? Of course we know very well that around that bend it looks very much as it does right here. Nevertheless, there's no denying the impulse to see. So we shoulder the packs, pick up a deer or aoudad trail on the east side shelf, and go have a look.

"It is more of the same, but somehow different, too. After half a mile along a ledge 30 ft. above the river we reach a beautiful open meadow with scattered piñons and junipers. It is an extraordinary year for piñon nuts, and this spot is a scrub jay's paradise. Across the meadow the river alternates between slow green pools and quick, rippling falls. At this stage the Canadian is not floatable, although it could be traversed by canyoneering, a sort of scrambling technique combining floating and hiking. Here a rock slide down the west wall has created a lane of ponderosa pines all the way down to the canyon floor. Ponderosas stick to the rims of the Canadian Gorge, but here's a big old lone pine, perhaps swept down as a seedling, looking a little wistful among the piñons and junipers. On the Llano above, between here and Wagon Mound, the Canyon Colorado Equid Sanctuary, a 6,000-acre ranch owned by William Gruenerwald, is dedicated to saving true wild horses and asses, like the Asiatic Przhevalski's horse. We keep expecting to see zebras peering over the rim. But all we see up there are circling hawks.

"We push through a salt-cedar thicket, mosquitos rising from every limb, spot a grove of wild china trees. Bill says the Kiowas he knows in Oklahoma seek them out for tipi stakes. Under a piñon on the W. slope we munch on dried fruit, scribble in our journals, swig Canadian River water boiled this morning—drinkable, but with a definite calcium taste—and try to figure out where we are from the topo maps.

"It is, perhaps, 11 A.M., and we step out into a gorge that is beginning to glare. We cross the river, the snowmelt water almost shockingly cold, at the prettiest rapids we've seen on the river. Jurassic sandstone encircles the lower walls of a round basin. Against a sheer wall, on the far side of a deep pool, Rocky Mountain maples are starting to put on some color. The entire river is deep here, forces us by its turns against the walls into a neck-deep crossing, our packs and cameras held high overhead. And now, surprisingly, the Jurassic pitches downward as if it's cross-bedded, forms staircase shelves over which the river pours.

"We stop at the mouth of a large side canyon where a grove of box elders provides a deep shade and look at the maps. A pair of rock pinnacles stands above the big rapids here, and they fix our position. This, not the point where we had camped, is the mouth of Cañon Mesteño. It's a little unsettling. Here it is noon, already, and we have at least 14 miles left to do. What the hell. Bob Marshall used to do such stuff for prebreakfast warm-ups. Of course, as Bill points out, Bob Marshall died from exhaustion at 38.

"It grows hot. We quicken our pace, passing isolated, mournfully lonely rock ruins. We slip through the last sandstone chute, one we thought we might have to climb around, at water level. A faint jeep trail appears along the river and we welcome it, the brisk, exhilarating walk of the morning turning a little wooden and mechanical in the hot glare of midday. Atop the blocky, Dakota rimrock, 800 ft. above us, pines groan and whip in the wind. But the deep gorge is still. Black grasshoppers with red wings rise from the grass. More box elders, and then a side canyon with a pinnacled mesa that must be Cañon Emplazado.

"'Be damned. Look at this.' Bill, purple bandanna stretched over his head, sweat dripping from that long, splendid nose, is inspecting a rusted sign. Faded white lettering warns us: Resettlement Administration, No Trespassing. But stretching away in every direction is remote, wild country, low, collapsed rock ruins the only evidence that the fed tried to relocate Exodusters here.

"The state archives over in Santa Fe tell the story. By 1925 or thereabouts almost 8,000 homesteaders had taken up Llano land in Harding and Mora counties on either side of the gorge. Within a decade the Dust Bowl hit. By 1935 one of the Resettlement Administration's projects was to relocate some 24,000 Texas and 9,600 New Mexico settlers. Someone had the idea of emulating the Hispanic settlement pattern and a Mills Land Use Adjustment Project was hatched to move ranchers from the gorge to the Llano and farmers into the gorge. The land exchanges were made. Homesteads were started, irrigation ditches dug. Then the Canadian River, whose normal flow through the gorge is about 5,000–10,000 acre-feet a day, sent 50,000 acre-feet surging through the sandstone chutes, and not just once. The homesteaders abandoned their tracts; even the old New Mexican town of Armenta, between here and Sabinoso, was abandoned. By the early 1940s the fed's Dust Bowl land reacquisition program took back the upper gorge and most of the Llano on the east side of it. In 1960, when the National Grassland system was established, it became the core of the Kiowa National Grasslands, was reseeded in native grasses. It took 20 years, but the wild finally reclaimed this country. Except for the rusted signs and the ruins of dust-blown, flood-washed dreams.

"We're now into the last haul—about 3 miles—and it is late afternoon, the wind up, temperature over 80, our metabolisms down and dying. Some trudging; long silences. We've reached the stage where the telling will be more fun than the doing, as Bill says. Again and again we wade the shallows of the river, lop off an occasional bend where the stream bed folds back upon itself. The canyon's geology continues the same as before, although those car-sized Cretaceous blocks at the rim are becoming house-sized. Here is a rock-slab grave and more ruins on either side of the Jeep trail. Desensitized by now, we are able to manage only marginal incredulity at how lonely and isolated from the world these homesteads would have been.

"Last swing in the gorge before the Roy/Wagon Mound highway. A big, thickly forested canyon—Cañon Blanco—comes in from the NE,

Specters of a ghost canyon:
TOP, Ruins of the old Mills Hotel;
BOTTOM, Rusting sign from the brief relocation experiment of the Dust Bowl era.

carrying a strong flow of water from numerous seep springs far up it. To the SE is another large stream flowing out of Beaver Canyon. Below where it comes in we climb down to the river, strip down, and where the Canadian has cut a sleeve into the sandstone we perform the old trapper ritual of rinsing off in the crick before hitting the settlements."

THE FOLKS in Ricardo's Bar that evening were satisfyingly daunted at the ambitiousness of our hike, maybe because by the time we got there many of them lacked the ambition to change barstools. Katie arrived soon after we did, and between ingestions of tequila and various other uncontrolled substances the locals regaled us with local stories of dust and floods and population busts, down to fewer than 1,200 people in all of Harding County today. (I later checked the census figures. Interesting. Neighboring Mora County lost 3,000 people during the 1920s but held on during the 1930s. Harding County had 4,421 people in 1930 and still had 4,374 in 1940. They survived the Dust Bowl. But discoveries of carbon dioxide fields failed to bring in an influx of people, and the underground aquifer that made other regions on the Llanos boom after 1940 was thin here; less than 15,000 acres were ever irrigated from it. Then came the floods of the 1940s and another drought in the 1950s.

Reclaimed by the wild: where farmsteads and ranches once stood, buffalo grass and chollas now occupy the Kiowa National Grasslands of the upper Canadian River Gorge.

It was too much. Small towns like Mills disappeared. As the country haired over again, ranching replaced farming. By 1970 Harding County had just three towns and 1,348 people and dropping).

Watching the world slowly recede from their borders doesn't seem to affect the residents of Harding County overmuch. At Ricardo's Bill was lured into an escalating world championship 8-ball tournament, whose stakes went even higher when a pair of young New Mexican girls challenged the table. Lewis, a quiet cowboy, kept buying drinks and asking Katie, "What's anthropophagy mean, anyhow?" Texas Tech and the University of New Mexico were thick into their interstate football rivalry in Albuquerque that night, and a radio was on. No one seemed to be listening. When Mike, the bartender, mentioned that Tech had won, someone belched. There were shouts from the poolroom, but it turned out they were for the 8-ball challengers.

Next morning Katie and I drove in from our camp on the rim of the gorge to look for Bill, who had taken his chances in town. A small crowd was gathered in a residential yard across from the city park where we were to rendezvous. Not spotting Bill in the park, we stopped to ask if anyone had seen him. The gathering—a Sunday morning outdoor mass?—parted as we walked up, and there was Bill, sound asleep, the contents of his pack strewn around him. The gathering sharpened. It included the mayor of Roy, the local constable, the overwrought citizen who had awakened to find this strange personage asleep on his lawn. While we reluctantly acknowledged our connection to this disturbance of the local peace, Bill opened an eye, staggered to his feet, almost succeeded in shaking hands with the mayor. It was a clear case of mistaken identity. In the dark of night he had taken the citizen's rather weedy yard for the park, and no harm intended. The mayor, an understanding fellow and up for re-election, was all for smoothing things over and allowed the possibility that Bill could have hiked the Canadian Gorge but gotten lost in downtown Roy.

FROM THE EDGE of the darkening gorge I watch sunlight perform its last theatrics atop the Sangre de Cristo range. The temperature is dropping fast now. It's time to haul this firewood back to Yellow House, four hours over those extending-into-infinity plains highways. I start the truck, fumble around for a tape to stick in the player. Townes Van Zandt is singing about a woman who fits just like his guitar when I pull onto the highway, but the imagery strikes me as uncomfortable and I turn the volume down.

THE FARMERS who settled the High Plains in the teens and twenties were dryland farmers. But the existence of the Ogallala Aquifer had been known since the 1850s, when Capt. John Pope had experimented with water wells in eastern New Mexico under government aegis. Pope's experiences initiated a maddening three-quarters of a century of mechanical failure in getting at the vast freshwater lake saturating the Llanos subsurface. But the Dust Bowl, with millions in loans available from the government's farm programs, acted as a catalyst to a resolution of the problem. In 1930 there were just 170 wells on the Llanos; by 1940 there were 2,180; by 1957, 42,225; by now, more than 70,000. In the half century between 1930 and 1980, irrigated acreage on the Llanos went from less than 35,000 to well over 6 million acres. This mining of the aquifer did nothing to stop massive dust erosion when the droughts of the 1950s and 1970s struck the Southern Plains, but in combination with the techniques of scientific agronomy as taught by the Soil Conservation Service and the state agricultural extension services, it did create a booming irrigation empire out of land Randolph Marcy had once called the "Zahara of North America." Who can blame plains people if they assumed their situation was a long-term adaptation, like that of the Archaics, rather than a too-narrow specialization with a meteoric but brief life-span, like that of the Clovis hunters?

Because, of course, the Ogallala Aquifer is fossil water, a lake whose original Rocky Mountain sources have been cut off by the Pecos and Canadian rivers, whose only source of replenishment now is through seepage from playa lakes. And 70,000 commercial water wells, each pumping 500 to 1,000 gallons a day, can suck the life out of even such a vast lake as the Ogallala at an astonishing rate. After five decades and a 250-to-300-foot drop in the aquifer level, farmers whose American roots have taught them not to accept limits, whose Texas culture balks at any kind of government controls to enforce conservation, and who often haven't believed in geology anyway are having to face the cruel truth.

It is bitter medicine to swallow, and not surprisingly it has produced a lot of denial. Some of the local underground water conservation districts have made some headway in stabilizing aquifer drawdown on the Southern Plains, and Wayne Wyatt, the director of District #1 in Lubbock, has fashioned a career preaching against the doom-and-gloom predictions.

Yet it is rationally impossible to ignore the fact of limits, and it would seem that the trend toward forcing recognition of that is already underway. A measure of the desperation lies in the farmers' insistence that Southern Plains agriculture is too important to allow to die. Texas ought to outdo California and construct an immense diversion project that will correct nature's error by pumping in water from somewhere else. Folks downstate in Texas and in the rest of the country either ignore or are ignorant of the situation, however, or almost seem to take a perverse delight in watching Webb's prediction about the desert come true. Meantime, faced with near prohibitive pumping costs as the aquifer level has declined, Southern Plains farmers have returned more than a million acres to dryland during the past decade. A 1985 study indicated that most of the remaining irrigating farmers would rather quit than return to dryland farming, and evidently this is what they're doing. Between 1980 and 1986 five of the nine counties in the heavily irrigated area around Lubbock suffered Dust Bowl–size population losses (up to 11.5 percent in half a decade) as farmers finally gave up.

Most modern American farmers cannot think in terms of a reduction in scale. The subsistence approach that keeps those New Mexican villages alive on the land after more than a century and a half would be unthinkable for the twentieth-century Texas farmer. And it is, of course, the lack of the opulent American lifestyle in favor of simplicity and continued ties with the land that is the secret of places like Sabinoso.

In effect, it seems that the present culture's life expectancy may be even shorter than that of the Clovis hunters—perhaps not much more than a century. Like that of the Comanches, whose cultural adaptation was never allowed to play out, the natural evolution of the present farming culture is not completed. Except, maybe, along the Canadian Gorge in New Mexico.

From the standpoint of Geology and scenery, Palo Duro is well worthy of being made into a national monument. It is the most spectacular canyon, carved by erosion, anywhere on the Great Plains of North America.

—Dr. Charles Gould, National Park Service, 1938

We believe that over the next generation the plains will, as a result of the largest, longest-running agricultural and environmental miscalculation in American history, become almost totally depopulated.

—Geographers Deborah and Frank Popper, 1987

Chapter 8

The Mythic, the Sacred, the Preservable

IF THE LESSON of history is change, and that of science to show consequence, what is the lesson of nature for human societies? It must be adaptation. Adapt or be banished from the play, say the plains. Adapt, or the rocks will seal your bones and nothing else will endure.

You get a totally different civilization and a totally different way of living according to whether your myth presents nature as fallen or whether nature is in itself a manifestation of divinity.
—Joseph Campbell, *The Power of Myth*

A few months ago, late in the winter, I drove down to the Double Mountain Fork Canyon not long after a heavy snowfall that had stuck for almost a week had finally melted. Patches of snow still gleamed whitely beneath the heavy juniper growth on the south wall, but the badlands section where Tule and I had baked the summer before was already clear of snow, wet patches of clay the only evidence still left. The day was bright, the winter sunlight dazzling on the red cliffs. I was moody, depressed that day. Tule was no longer with me, having died in a bizarre accident. Katie had moved on, to whatever fate awaited her in New Mexico. The sun, the vermilion slickrock, the pungence of junipers, the weird forms of the badlands were exactly what I thought I needed.

As soon as I saw the badlands I knew something was wrong. Something . . . was missing, and it took a long moment before I realized what.

It was my favorite dinosaur hoodoo. It was gone. Not the massive red clay mound, of course, but the hoodoo itself. As I walked toward it I conjured all sorts of explanations: some idiot had blasted it off with dynamite, the goddamned machine-mad dirt bikers had cut yet another trail and somehow weakened it, high school kids from Post had pushed it off as part of a prank or dare.

But when I climbed the mound there was no evidence at all of human tampering—just an oblong cavity where the pedestal had stood, an avalanche of debris down the slope, and the hoodoo's sandstone cap lying there, like any number of similar slabs scattered across the surface of

OPPOSITE PAGE:
The Tule Canyon Narrows, once one of the most sacred, now perhaps the most preservable place on the Southern Plains.

the badlands. What had happened was clear. The hoodoo's time had simply come. The wet snow had weakened the base of the pedestal, and the strong chinook winds that had followed the norther had toppled it. Geological time had intersected human time.

Such an event should not have surprised me. I've tried for years to see time in these canyons. I've concentrated, for instance, on seeing Turkey Mesa out my bedroom door as a frozen wave. A few thousands of years ago it was part of the canyon wall. Narrow the eyes and use imagination to speed time and I have fancied I can see it whittling to a butte, crumbling to a rounded hill decorated with slabs of the massive caliche that now caps it. Behind the tipi is a gigantic white boulder the size of a Volkswagen; 100 feet directly above it the corresponding Volkswagen-sized notch in the cliff. I've found bison skulls, Archaic spearpoints in these canyons. Time is visible in the landscape here, just as adaptation, or the lack of it, is visible in the historical and archaeological records.

Ah, you say, but mostly these journeys have dealt with past time, much easier to see and read. What of time to come?

At least for the moment, project the trends forward. The hoodoos will continue to fall. New ones will take their places. The canyons will continue migrating upstream, their bottom ends growing wider all the time, until eventually those on the Llano Estacado will demolish the plateau completely, leaving a mesa-badlands and broad-valley country, like that already carved by the Pecos and Canadian rivers, where the Llano Estacado once stood. Say, another million years? It should be something to see. But since the average mammal species only lasts two or three million, will any of us be around to see it?

"I AM NOT SO SURE but that the prairies and plains, while less stunning at first sight, last longer, fill the esthetic sense fuller, precede all the rest, and make North America's characteristic landscape." So wrote Walt Whitman in *Leaves of Grass.* He was one of the few people (along with Georgia O'Keeffe) in the United States to possess such an opinion. We Americans have not so much had our aesthetic sense filled by the plains as we have denied that the plains are aesthetic. In fact, grasslands and the ecosystems occurring in grasslands are among the most underprotected natural areas worldwide. Interesting response from a species whose early evolution owes much to the plains of Africa.

We North Americans gave the national park concept to the world. Yet, in a fine irony, although the first visionary call for a "great national park" by artist George Catlin in 1832 was for a park on the Great Plains of the American West, the plains is the most underrepresented biotic province in the American park system. Only in the past decade, with the creation of Canada's Grassland National Park (1981) and the upgrading of Theodore Roosevelt Memorial Park (North Dakota) and Badlands National Monument (South Dakota) to full national park status (both in 1978), have large, wild parks been set aside on the Great Plains. Although the Southern Plains canyonlands are more scenic and more geologically important than these northern parks, not a single national park or monument exists here. That requires some kind of explanation.

The obvious candidates have always been the Red River canyons, Palo Duro and Tule in particular. And interestingly enough, early in the history of the National Park Service (NPS), specifically from the 1920s through the 1940s, there was a running debate within the service over the merits of creating the "National Park of the Plains" around Palo Duro Canyon. It's an intriguing story, one that has lain buried in the NPS papers in the National Archives for half a century. The conclusion oughtn't to wait: Palo Duro would be a national park at least ten times larger than the current state park if it hadn't been located in the state of Texas. Maybe not initially, as a result of the park service's evaluation of it in the early 1930s, but it would have become that large, vastly multiplying its importance.

Palo Duro, because of its role in western history, ranked alongside the Black Hills as the most famous Great Plains landscapes in the early part of this century. But unlike the Black Hills, visible from a distance, Palo Duro was tucked away in the basement of the plains. A certain mystery prevailed about it. Texas, of course, had privatized almost all its plains country, and because of their water, the canyons had gone into private hands quickly. Five large landowners, the biggest the JA Ranch, had cut off public access to Palo Duro. Rumors about the scenes in its depths were passed around the plains. To convey an idea of how much curiosity there was then, how hungry people were just to see Palo Duro Canyon, when a landowner in 1932 announced a one day open-house drive on unimproved ranch roads, 25,000 people from fourteen states inundated him.

There had been a mounting clamor over access to the canyon and the state or federal acquisition of it since Theodore Roosevelt had set up his conservation program of national forests, parks, monuments, and wildlife refuges. In 1908 Texas Congressman John Stephens introduced a bill calling for a Palo Duro "national forest reserve and park," an idea that must have confused everyone. The bill died in committee. Eight years later, after the park service had been created under legendary director Steven Mather, Palo Duro was brought up as Texas' first candidate for a national park by Senator Morris Sheppard. Over the next two decades it was discussed frequently in Washington, and the canyon was twice visited and evaluated by park service personnel.

History should derive its interest not only from its explanation of how things have come to be but also from the opportunities it allows for Monday morning quarterbacking. We are supposed to draw decision-making insight from history. Yet western history, at least, is read by most popular consumers more as a hobby than an intellectual exercise. The only thing the majority of readers of western history second-guess these days is the treatment of Indians, not because they are so much empathetic as because they recognize intuitively that we substituted something soulless for something fine, wild, and free.

The sequence that left us with the tiny, 16,400-acre state park in Palo Duro is one history that ought to be second-guessed. Under Mather and his successor, Horace Albright, a strict set of criteria was developed for additions to the parks and monuments system. The buzzword then was "monumentalism." The NPS at the time knew little and cared less about ecology, diversity, or preserving representative ecosystems. It wanted its parks gigantic and grand. Yellowstone and Yosemite it had inherited. Now, in the 1920s, it set about completing a "crown jewels" natural showcase system. The Grand Canyon was upgraded from national monument to full park status in 1919, and Zion National Park, created around the stupendous canyons of the Virgin River in Utah, was added that same year. Bryce Canyon joined this group in 1928. None of the canyons or badlands on the Great Plains had vital statistics that looked as impressive in this crowd. When Roger Toll, the chief investigator for the NPS, examined the South Dakota Badlands in 1928, he pronounced them "surpassed in grandeur, beauty and interest by the Grand Canyon National Park and by Bryce National Park . . ." But since 60 percent of the Badlands had not been homesteaded, and since South Dakota promised to buy and transfer to the NPS 90 percent of the private lands, the service decided to make the Badlands a national monument of some 250,000 acres.

Mather had encouraged a state parks program, and Texas, under Gov. Pat Neff, had responded by asking for land donations and getting in return some of the worst land in the state. By the early thirties, there were some three dozen Texas state parks. And the crown jewel of the Texas parks was its brand new (1933) 15,000-acre Palo Duro Canyon State Park. The creation of this park satisfied the immediate desire for access; among some it still does. But settling for a few slices of a very large loaf had important negative consequences for the eventual creation of a major national park on the Southern Plains. Those interested in conservation history might derive something of value by noting how this compromise turned out.

For one thing, creation of the state park took the edge off pressure and enthusiasm for national designation. For another, it put NPS officials in a difficult position. They would set a discouraging precedent for the state parks movement if they proceeded to absorb the best state parks into the national system. They also had begun to suspect that they had been stung when they aided Texas efforts to finance the acreage for the state park in Palo Duro. One Fred Emery, a real estate salesman from Chicago, had, through foreclosure, acquired 18,700 acres of the most scenic part of the canyon. Emery's land became the state park, and in the process he fast-talked a deal with the State Parks Board that gave him nearly twenty-five dollars an acre for land that normally sold for five. Emery kept bobbing up like some sort of money-junkie for the next decade.

Despite all this, the NPS sent Roger Toll (as part of the Texas tour that led to the eventual creation of Big Bend National Park) to evaluate the Palo Duro system as the potential "National Park of the Plains" in early 1934. Toll spent four days in the Red River canyonlands, starting at the Tule Narrows and ending at Dreamland Falls. The proposal under consideration was a one-million-acre plains park stretching from the falls all the way down to today's Caprock Canyons State Park. Fifty miles of Palo Duro, plus the Tule gorge and the canyons of the Little

Considered variously for national park and national monument status, neither of these places is part of the public lands today. Both are top candidates for plains wilderness protection.
TOP, *Tule Canyon;*
BOTTOM, *Palo Duro Canyon.*

Red, would have been protected, along with a strip of high plain on either side so that pronghorn and bison could be reintroduced.

Alas. Son of a bitch! Toll did regard the Red River canyonlands as scenically superior to the badlands that were already becoming national monuments on the Northern Plains. But based on monumentalist criteria, they were not the equal of the canyons of the Colorado Plateau. Palo Duro's characteristics were "so much like those of the Grand Canyon," one NPS official wrote, that "the Palo Duro as a national park would be a 'tail to a kite.'" Also, in contrast to Palo Duro, a well-lubricated acquisition existed in Big Bend where land averaged about two dollars an acre (and sometimes went for a penny an acre) as local ranchers sought to unload desert grasslands they had destroyed by overgrazing.

Bad enough, but it gets worse. Economic turmoil had the Southern Plains in trouble. The JA Ranch seemed seriously intent in these years on unloading its part of Palo Duro, at one point offering the ranch to the newly oil-rich Osage Nation, at another proposing that the U.S. Defense Department acquire Palo Duro as a bombing range. Homesteading had ended, crushed by the Dust Bowl. Some wanted the federal government to reaquire the entire Great Plains.

In 1938 the park service decided to do a new evaluation of Palo Duro, this time far more thoroughly (a team of eight experts) but with the more modest idea of a 135,000-acre national monument extending from the falls to the Claude Crossing. The investigation was thorough and overwhelmingly in favor of the monument on the part of the field team. The noted geologist and NPS advisor, Herman Bumpas, added his arguments. Bumpas emphasized the geological value of the canyon as a vital link in the service's new theme of public education. He argued that, for tourists driving west, Palo Duro could play the role of the "First Chapter of Genesis" and that, for the public to grasp fully western geology, seeing Palo Duro was essential since its geologic strata ended exactly where those of the Grand Canyon began. Charles Gould, one of the investigators, put it directly: "It is the most spectacular canyon, carved by erosion, anywhere on the Great Plains of North America."

The Washington NPS staff was divided on the recommendation, essentially over one issue: Palo Duro was in Texas, where all the land was privately owned, and the 135,000 acres was assessed at nearly $560,000. And now came Fred Emery, offering to sell his remaining 3,000 acres "NOW . . . in a spirit of cooperation . . . to see this wonderful and colorful scenic and recreational and historical . . . game preserve and bird sanctuary . . . nationally and internationally known [developed]" for a mere $475,000.

That was about it. When Senator Sheppard inquired about the status of the project in 1940, he was told by NPS officials that "the Department probably would be willing to recommend the establishment as a national monument of approximately 135,000 acres of land . . . if the necessary area could be acquired." Private funds had been raised to acquire lands to create parks, such as Arcadia in Maine, Shenandoah in West Virginia, and Great Smokey Mountains in South Carolina. Texans like Senator Sheppard and Amarillo's Guy Carlander and Phebe

Warner, among others, tried to keep national acquisition of Palo Duro alive, but the Great Depression and the war were too diverting. And from Texas, with all its oil millionaires, there were no offers of help in buying the land.

In 1978, with the passage of the Omnibus Park Bill of that year, both of the major national monuments created on the Northern Plains during the 1930s and 1940s became full-fledged national parks.

THE MODERN INHABITANTS of the Llanos are descendants of recent pioneers, and their mythological heroes are pioneer heroes—Indian fighters and buffalo hunters, cattle ranchers and sodbusting farmers. These kinds of mythic heroes, unfortunately, are becoming rusty relics in the modern West, with its worries about droughts, poisoned and depleted aquifers, creeping desertification and the Greenhouse Effect, and the shrinkage of wilderness and of biological and cultural diversity. Yet modern Southern Plains culture is so recent that it hasn't yet had the time to develop a mythology more appropriate to the circumstances it finds itself in today.

Direct expressions of landscape mythology, like the outdoor dramas and musicals put on at Caprock Amphitheatre State Park, Palo Duro's summer musical, *Texas*, Blanco Canyon's *God's Country*, and (my favorite) Shallowater's *As the Windmill Turns*, tend to be naive horse operas that resonate nineteenth-century ancestor worship rather than twentieth- and twenty-first century significance. For similar reasons, many of the best-known Southern Plains historians have written as uncritical, symbolic frontiersmen. Even regional high literature like Elmer Kelton's *The Time It Never Rained* and *The Day the Cowboys Quit* is most often a celebration of the heroism of mythic frontier types. I wonder if the Clovis hunters who likewise committed wholesale alteration of plains ecology weren't worshipped as giant slayers by their descendants.

What is being celebrated is the process whereby the land was geometrically carved and tamed, yielding to the human will, the Hispanics pushed back to New Mexico, and the Indians driven to Oklahoma. The wild, pagan, sensuous men and women have been banished, the land de-buffaloed and de-wolved and de-grassed, and civilization sits in the catbird seat, with the symbols of wilderness subjugation pinned on its chest like a butcher-bird pins its insects on a barbed-wire fence. Wilderness histories of the plains sing of buffalo hunting on autumn mornings, of mountain men and Comanchero traders, of thronging congregations of wildlife, of making love beside the fire, of untrammeled wilderness and freedom. But look at the photographs in the last chapters of the texts: sorghum, wheat, and cotton fields, feedlots, swooping planes spraying poison across the landscape, vapor lamps, brick houses that look interchangeable. For this we wax lyrical over the pioneers?

Not all the high art of the Southern Plains is misty-eyed over these myths. Larry McMurtry's early novels aren't, in contrast to *Lonesome Dove*, which is, sort of. Neither are the novels by Max Crawford, or books by people like Donald Worster, Marc Reisner, or Charles Bowden, or Walter Prescott Webb's later essays, or Alexandre Hogue's Dust Bowl paintings. But these people have all been either outsiders or expatriates. And their works don't go down well with the locals. Amarillo has never forgiven Hogue; nor Lubbock, Webb; and Sweetwater still vilifies Dorothy Scarborough for writing *The Wind.* Not only do the plains people cling to their myths, they also make it uncomfortable for anyone to write or paint in a gutsy, critical way about them. But, if wild nature is to survive and even thrive and be meaningful on the plains, we need to create new myths, or need somehow to reshape the old ones so that what is celebrated is not so consistently the grinding of nature and of diversity under the boot.

That was a process, after all, about which some of the pioneers themselves expressed ambivalence, even regret. "The beauties and blessings of civilization are very largely a myth," old trader and hunter James Mead wrote in his memoirs. "The freedom and beauty . . . of the plains are a thing of the past; nothing now . . . remains but dull, plodding labor."

THE CONCEPT of sacred place, of power spots in the landscape, is not part of the dominant culture's landscape mythology today, though the idea is an old one that even goes back to pre-Christian times in Europe. The Greeks, for example, had their sacred groves and their power spots where they could commune with certain gods. The Llanos canyonlands country was, and still is, full of such places, although our removal of the natives was so rapid that few of their sacred places have been remembered. Among these are the Medicine Mounds near the confluence of the Pease and Red rivers. The Double Mountains down on the Brazos probably were another. I climbed them a few years ago. From the west peak, the view of the Callahan Divide 50 miles to the south, of the Mackenzie Mountains and other detached mesas of the Llano Estacado to the west, and, in between, of a vast glowing panorama of rolling plains encircling the horizon is stupendous. The east peak of the Double Mountains was a perceptive choice for the vision-quest scene in Kelton's *The Wolf and the Buffalo.*

Deep in the canyonlands themselves, unquestionably were others. The Tule Narrows must have been a sacred place for every culture that has ever occupied the Southern Plains. Ancient foot- and handholds lead down the sandstone cliffs to the big spring, the most stirring descent into the Earth Mother anywhere on the Llanos. When the aquifer was full (the locals say that, once, you could hear the falls from the rim, three-quarters of a mile away), Lingos Falls probably was the home of a powerful water spirit. Roaring Springs may have been, too. Flint quarries in Quitaque and Tule, and certainly Alibates Quarry in the Canadian Breaks, were medicine places. Some of the freestanding mesas in Yellow House and Blanco canyons seem to have been singled out by some of the early mapmakers as special places, and Tucumcari, Quitaque, and Muchaque peaks were distinctive enough to have preserved their Indian names. In such flat country, all these elevated peaks were probably the scenes of hundreds of vision and power quests over the

OPPOSITE PAGE:
Moonrise over the Kiowa National Grasslands, eastern New Mexico

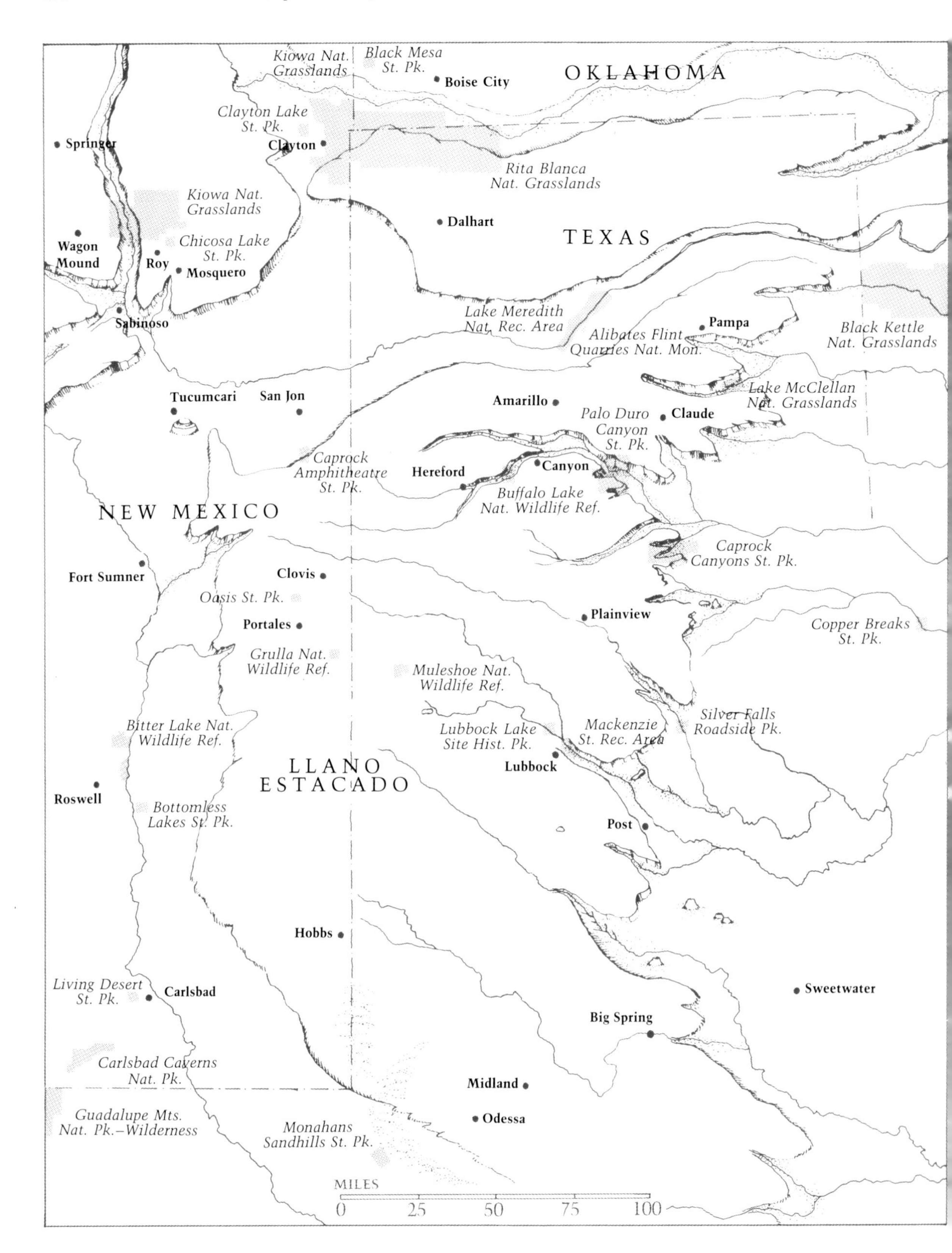

Present Preserved Wildlands and Other Protected Sites in the Modern Southern High Plains

centuries. High overlooks across North America were power places in Indian cultures.

The Comanches, Kiowas, and Cheyennes have preserved sizeable bodies of their mythologies and some of them are landscape associative. One of the most persistent and widespread mythological stories on the Great Plains was of the annual emergence of the vast bison herds from the depths of a great canyon on the Llano Estacado. I've never seen any particular canyon singled out by this story, but it's a Palo Duro–sized mythology. And one that, eerily, may be borne out by paleontology, since by far the largest number of early specimens of *Bison bison* have been unearthed in the Red River–canyonlands region. It could be that 9,000 years ago the Palo Duro country was the scene of the evolutionary emergence of the modern bison. If so, this might be the oldest remembered human mythology of the plains.

The mythology of sacred places in the Southern Plains canyonlands has mostly been lost on modern inhabitants. Old-time Christianity regarded such nature shrines as the abodes of Satan. Modern Christianity's ideas about wilderness have changed as society has evolved, and today even church camps are located in canyons like South Cita and Blanco—hardly the same thing as nature shrines to spirit of place, of course. But the sacred places are still here, secreted away in the canyonlands, waiting rediscovery by the sensitive or by Indians with remembered traditions, waiting appreciation by some new cultural development on the plains.

Wallace Stegner, in *The American West as Living Space,* believes that it will be the task of artists who stay and immerse themselves in the American West to push contemporary cultures in the direction of new myths based on bioregionalism and rootedness in nature. As Joseph Campbell points out in *The Power of Myth,* artists in modern societies are the cultural equivalent of shamans in primary ones, and shamans were the principal myth-givers to early humankind.

Perhaps it will work this way. Certainly we have reached the stage in human evolution where we need new myths, new mythic heroes. The universal heroes who have shown us the way to spiritual acceptance of death as integral to life, and who have instructed us in how to subdue the earth, must now make room for the new mythic hero, whose function and lesson will focus on saving the earth and recentering us in a new, civilized animalness.

But painters and writers with unorthodox views have always gotten into trouble on the plains. And literature and art, like science and reason, may be insufficient anyway to summon a nature ethic for the West. Nietzsche was right: history does seem to show that religion is the only motivation capable of really causing humans to act reverently toward anything. But what religion? The great 1960s debate over Christianity's stance toward nature was never resolved, but it was generally agreed that there is a problem in an anthropocentric religion that places humans in a superior position to the rest of creation because they alone are made in the image of God. Unfortunately, the old nature religions, despite their views of the earth and all creation as sacred and of humans as fellows of all other creatures, were not able to grasp cause-effect relationships in nature clearly enough to avoid disasters.

Ecological science is beginning to do that well, but for most people, rational explanations spoil the magical and the supranatural. And science, after all, gave us concepts like animate and inanimate, which reduced the universe to the material and the measurable.

Religions are born in crises. Any new nature religion that combines aspects of Christian mysticism like John Muir's, pagan nature worship, and ecological science in a creative and workable mix is not likely to evolve until grave ecological crises emerge that we cannot avoid. Then, mayhap, we will have our new myths and our new heroes.

In any case, creating new myths is a serious challenge for artists who would attempt this act of modern shamanism. A rediscovery of the old Southern Plains *sanctum sanctorum* by Native American painters and writers may well be critical to the contemporary culture's chances for a truly symbiotic relationship with this landscape.

IT ALL COULD HAVE BEEN very different on the Southern Plains had that million-acre Palo Duro National Park become a reality back in the thirties, or if a Palo Duro National Monument had gone on to achieve park status in 1978. That didn't happen but, unquestionably, would have produced a ripple of environmental awareness across the Llanos if it had. For one thing, the park would have kept creative people on the Southern Plains and drawn others here, which the present Llanos culture has always had trouble doing. Something like a large wilderness park seems the place to start. Something like it may be the one best chance the present culture has if the plains are not to become again a national ecological disaster area.

Nature preservation in, and public access to the Llanos canyonlands is paltry. About 15 linear miles of the upper Canadian Gorge are in the Kiowa National Grasslands, managed at present for grazing and a bit of recreation. One tiny New Mexico state park, Caprock Amphitheatre, is located in the Western Caprock Escarpment near San Jon. You can throw a rock across what escarpment it preserves. On the Texas side of the Llanos, Caprock Canyons State Park adds about 14,000 acres to Palo Duro park's 16,400, but 30,400 acres of land, a good bit of it developed for tourists, preserves a pathetic 3 percent of the vast park the National Park Service once evaluated. Canyonland diversity and total acreage will go up if and when the proposed new state park in Blanco Canyon is added. The reservoir "parks" in Yellow House, Tule, and McClellan Creek canyons, however, add little or nothing to nature preservation in the canyonlands. Neither does tiny Silver Falls Roadside park in Blanco. The new Lubbock Lake Site Historical Park in Yellow House Draw is an important state addition archaeologically but not otherwise. Although there are other state and federally protected areas on the Southern Plains, including about 262,000 acres of National Grasslands, none is in the scenic canyonlands. And Texas' Copper Breaks State Park is NOT a canyonlands park, despite the park commission's claim.

This is a pretty sad commentary on the state of environmental awareness and preservation of the only major scenic wildlands on the Llanos, and it means that presently the most spectacular and, scientifically and spiritually, the most important country on the plains is serving no higher function than poor stock range. Given the demographic trends on the rural plains, though, the opportunity to do something noteworthy that would turn the Southern Plains around is here as it hasn't been since the thirties.

A certain self-inflicted knife twisting exists in the lack of sophistication of plains culture toward environmental issues. Ideas like the value of wilderness to society and modern philosophical and theoretical debates over ecological limits, "deep ecology," and the land ethic are ignored by the plains media and do not exist for most of its present population. But some of those issues are going to be fought over the Great Plains in the coming decades. Already, in the face of a dwindling and increasingly polluted aquifer, rural stagnation and population decline, and the hovering threat (every spring that the rains come late a kind of mass unease sets in) of the Greenhouse Effect and dust bowls, radical suggestions that haven't been voiced since the thirties are being published and debated.

Bret Wallach, a MacArthur Fellow and geographer from the University of Oklahoma, is one expert who's making them. Wallach's proposal is straight from the New Deal. He wants the federal government to begin a program of acquisition of river lands on the plains to be made into either "national prairie parks" or new national grasslands; he seems unconcerned about the differences between those two types of public lands, but does mention that at late-1980s land values acquiring three million acres would run about half the cost of a single aircraft carrier. He has worked out a mechanism that would restore strips of the native plains through 15-year voluntary contracts with landowners, during which time the government would pay for crops or stock not raised, that would end with title to all but a 40-acre homestead vested with the federal government. Wallach realizes that to many ranchers and farmers such ideas "will offend their every instinct." He also realizes it will cost money. But, if they offered something exciting and wilderness-oriented by way of recreation, a vastly expanded national grasslands plus a series of plains parks could offer one dramatic method for diversifying and turning around the hopeless direction of plains culture.

Other ideas include one that makes Wallach's seem timid. Two Rutgers University geographers, in an essay they call "A Daring Proposal for Dealing with an Inevitable Disaster," believe that over the next generation, as a result of what they call "the largest, longest-running agricultural and environmental miscalculation in American history," plains culture will totally collapse, never mind the enormous subsidy investments poured into the region to buttress privatization and despite the recent Conservation Reserve Program for returning eroding farmlands to native grasses. Their future scenario projects an acceleration of the gradual impoverishment and depopulation that is presently afflicting the rural plains, especially in Texas and New Mexico, in the 1980s. With the countryside being deserted, a need will arise for a new Resettlement Administration. Eventually, what a few realized all along will strike with the force of revelation: the best and highest use of the plains is not as cotton or wheat fields or even cow pastures made superfluous by a changing American diet, but as short- and mixed-grass prairie with native grazers. According to Deborah and Frank Popper, the future holds forth a restored "Buffalo Commons" on the Great Plains.

Intellectuals like to play games with theories about how and why things happen. Walter Prescott Webb's environmental determinist hypothesis was one of those that seemed to offer an insight, but one that subsequent history has stood on its head: the plains adaptations turned out to be Rube Goldberg gimmicks to enable the new culture to avoid really adapting to living within the limits of plains ecology. The geographers have offered up a theory called possibilism to explain how cultures with different technologies and ways of "seeing" an environment choose how they live in it. Any number of possibilities exist, and choices—like those of southern pioneers to plant the Llano Estacado in cotton and midwestern settlers to plant it in wheat—are products of cultural filters.

A crisis is on its way for the Southern Plains. The present culture has not adapted, it has mined, and the end of the vein is in sight, in some places has already been reached. Those who argue for technological fixes plead that "something" will come along. Grapes, a friend tells me—wine grapes are the crop of the future on the Llano Estacado. But he hasn't considered that only 2,000 acres of grapes are presently producing enough for more than 50,000 cases of plains wines, and that the year 2000's projection of a paltry 10,000 acres of plains grapes will saturate the market, unless an earthquake shakes California off the continent by then. I would argue that the "something" was here in the first place.

It was the wilderness plains, which we couldn't see for wanting to recreate the East, or maybe the Old World, on a landscape where that was grossly inappropriate. Why not, piece by piece, put the wilderness back together—if we just leave, the land will naturally revert, but the process can be far faster if we help—and, like the Archaic cultures, adapt locally and imaginatively to it? We have to face the fact that we've botched it, that the century-long experiment to see if the plains could handle economic patterns like those that seem to work elsewhere is coming to an end, that aquifer drawdown and global warming are not going to be ignored into nonexistence. Loss of human population to these changes can be seen ecologically as a positive good; there may be fewer of us, but the quality of our lives can be extraordinarily richer.

The canyonlands are the key to the adaptive door, I think. It may take another half century, but we can develop a new culture based upon giving the plains wilderness a dramatic foothold and adapting to a recreation-tourism economy along the lines of Alaska's. Except what the Southern Plains will offer is the American West, which drew people from all over the world in the nineteenth century.

The scale of the start might be small. Texas Parks and Wildlife's current six-year plan (1988–1993) calls for the acquisition of 72,000 acres of new parkland, especially "natural areas of statewide significance." The first wildlands priority on the Southern Plains unquestionably is getting a state wilderness or nature preserve in Tule Canyon. The region begs for privately funded Nature Conservancy tracts (at present there is not a single one on the Llanos) protecting rare and threatened species and communities and more state parks stressing wilderness values in Los Lingos, Quitaque, Yellow House, and Double Mountain Fork canyons. A Caprock Escarpment national or state hiking and riding trail, extending from the Muchaque Valley to the Wolf Creek Valley, is a worthwhile project that could be started piecemeal and built up over a couple of decades.

Ultimately, the Southern Plains ought to put a heroic effort into getting the entire Palo Duro ecosystem established as the great nature preserve we failed to get in the 1930s, this time around not as a traditional western park but as a national interest wilderness like those created in the Alaska Lands Bill of 1980, with bison and antelope, mule deer and bands of wild mustangs, and cougars, bears, and plains lobo wolves all reintroduced. These native animals deserve to be restored; they have the same right to the plains that we do.

Such a Southern Plains wilderness could also implement special-use privileges for the Comanches, Kiowas, Apaches, and Oklahoma Cheyennes. The loss of Native American culture from the Llanos is one of the region's most foolish and costly pioneer legacies. Finally, from these protected corridors, we would be ready to expand the national grassland system across the plains, push it in the direction of native grasses and grazers, turn Lubbock and Amarillo, Las Vegas and Clovis and Boise City into outfitting and service towns for recreation and tourism. The Southern Plains could become the American Serengeti Plain once again.

It is too fantastic. It's a pipe dream. Of course it is. It's an act of the imagination on a par with taking a wild country and turning it into an agricultural empire in half a century. Except that it has a chance of lasting a lot longer and being a lot more interesting in the bargain.

SOME IMMEDIATE THREATS menace the canyonlands. A universal one is gravel and caliche mining. It goes on in most of the canyons that are near the larger cities and towns, destroying not only natural landforms and vegetation but also prize archaeological sites (since the mines are mostly on private lands, environmental impact statements aren't required). The Blackwater Draw site, one of the most famous archaeological sites in North America, has nearly been eaten away by gravel mining. Big chunks of Yellow House and Palo Duro canyons are bitten off. As I've elsewhere mentioned, in semiarid environments seriously disturbed vegetation takes half of a human lifetime to recover, even where surface contour reclamation has been done. And reclamation laws, in Texas particularly, are weak and often ignored.

People also tend to regard the canyons as holes to be filled with municipal trash or tornado debris or, worse, as potential bathtub reservoirs as plains culture casts about desperately for alternative sources of water to keep its present paradigm going. Building the already-existing lakes sacrificed extraordinary natural areas, like Dreamland Falls in Palo Duro (underneath Lake Tanglewood) and the upper Tule Canyon

Basin, an area of numerous waterfalls and great beauty (beneath Mackenzie Reservoir). McClellan Creek Canyon has mostly become an offering to oil development and a reservoir. Camping there is possible only for the hearing impaired or those (and some do exist in oil country) who have been behaviorally altered to enjoy the infernal creaking of pumping units.

Yellow House Canyon's string of treated sewer-water lakes in the draw above the canyon, in the park named for Lubbock's famous native son, Buddy Holly, is no doubt an improvement over the industrial dump grounds that were once there, although it might give pause to residents downstate to realize that the Brazos River now has its "headwaters" in flushing toilets in Lubbock. Two residential reservoirs, Buffalo Springs and Lake Ransom, are in the canyon itself, part of a scheme that once envisioned a lake also flooding the mile-wide valley where I live. (These lakes and the springs and groundwater in upper Yellow House have been polluted for the next century by the myopic Lubbock wastewater facility; Amarillo evidently intends the same fate for Palo Duro water, while it simultaneously urges Parks and Wildlife officials to stock the canyon's river and springs with rainbow trout. In fact, so many toxic storage tanks have been lately discovered to be leaking into the aquifer that all the springs and rivers draining out of the Ogallala Aquifer in West Texas may end up poisoned if something isn't done quickly.) The burg of Claude has long wanted to build a dam across the Prairie Dog Town Fork and flood Palo Duro below the Claude Crossing. The idea of a dam across the Tule Narrows makes my heart palpitate and icy sweat run down my sides, but it's been suggested.

Fortunately, many of the current and planned reservoirs—Conchas and Ute Creek on the Canadian, Greenbelt Reservoir on the Salt Fork of the Red, White River Reservoir, the Justiceburg–Yellow House projects, and Thomas Reservoir on Tobacco Creek—are downstream from the main canyons, where there is additional waterflow. Dams affect ecology both above and below them, though. Like the concrete faucets they are, they turn the rivers on and off like garden hoses. And losses occur. The Justiceburg Reservoir will inundate a dozen rock-art sites, one a Plains Indian representation of an American wagon train, beneath its pool. And there will be more bathtub reservoirs and more water pollution and mining, unless we get the canyonlands protected.

AS I WALKED OUT of the Double Mountain Fork Canyon in blue dying dusk last winter, I came to two conclusions. Namely, that not a single one of the canyons, or the mountains, or the plains of the American West gives a goddamn about us. Nature is not moral, not immoral; it's amoral. And entirely unconcerned with our petty cultures, whether they succeed or fail, establish glorious empires in its midst or recoil in tragedy and defeat when nature's structures fall apart around them. A volcano erupts, a hoodoo falls, a society overextends itself. Nature cares naught about our reaction one way or the other.

The second thing was this: all our "biocentric" concern about Nature is, scratch the surface only a little, in reality a self-centered concern about ourselves. Environmentalism may be, to paraphrase Edward Abbey, the late-flowering consciousness of the world mind—the earth's voice—but on a more mundane level it often seems nothing more than the anguished thrashing of a sentient species that has reached the outer rim and realized that the earth really is flat, of the smart monkey whose sticks won't fit together and won't reach the highest fruit. There are limits out there, beyond which Nature will respond, in its indifferent, amoral way, by squashing us. And not care how loud we holler. Any more than we've cared about the earth when we suck its aquifers dry or perform systematic genocide on the lifeforms it created to thrive in all its special places. Human morality is an evolutionary byproduct of the needs of a social species. And environmental ethics arises from those same needs, much as the old nature religions respected Nature because of our primal need to influence its forces.

Human cultures have reached this sort of ecological crisis before, now that I think about it. The Clovis hunters did: they came face to face with the rim of the world and probably ascribed it all to divine providence.

That's one reaction, of course. It takes the pressure off. But possibly, just maybe, while we're peering hopefully off the rim of the Southern Plains world we'll notice what a firm and potentially symbiotic grip our toes and the terra firma have on one another.

Selected Bibliography

A BIBLIOGRAPHY is an intrusion in a literary work, but this book seems to demand one and not merely to prove that I have not made all this up. Some of the conclusions presented here represent original research; I mention a considerable number of other authors' works and points of view; my own worldview has been shaped as much by what I have read as what I have experienced; and, most important, this bibliography can be used as a starting place by those who wish to delve further into the subject of humans and environment on the southern High Plains.

No list of maps is included among the entries of books, articles, reports, and private papers that follow. I ought to mention that, in rendering the names of canyons and rivers in this book, I have followed modern U.S. Geological Survey maps rather than current regional custom. For example, the official cartographic spellings of two of the Texas canyons are Yellow House (rather than "Yellowhouse") and Los Lingos (and not, as it is locally pronounced, "Linguish"). Consistent with the treatment of the word "tule" in the botanical literature, Tule Canyon has been spelled here without an accent mark over the *e*, although some early maps and one or two authors do so to emphasize the proper pronunciation: Too-lee.

INTRODUCTION AND 1. LAND OF THE INVERTED MOUNTAINS

Agenbroad, Larry. "New World Mammoth Distribution." In *Quaternary Extinctions*, ed. Paul Martin and Richard Klein. Tucson: University of Arizona Press, 1985.

Brand, John, ed. *Mesozoic & Cenozoic Geology of the Southern Llano Estacado.* Lubbock: Lubbock Geological Society, Department of Geosciences and ICASALS, 1974.

Collins, Michael. "A Review of Llano Estacado Archeology and Ethnohistory." *Plains Anthropologist* 19 (1974): 134–145.

Crosby, Alfred. *Ecological Imperialism: The Biological Expansion of Europe, 900–1900.* New York: Cambridge University Press, 1986.

Cummins, W. F. "Notes on the Geology of Northwest Texas." *Geological Survey of Texas,* 1893, 177–238.

Dillehay, Tom. "Late Quaternary Bison Population Changes on the Southern Plains." *Plains Anthropologist* 19 (1974): 180–196.

Gustavson, T. C., ed. *Geomorphology and Quaternary Stratigraphy of the Rolling Plains, Texas Panhandle.* Austin: Bureau of Economic Geology, University of Texas, 1986.

Hester, Thomas R., ed. *The Texas Archaic: A Symposium.* San Antonio: Center for Archaeological Research, University of Texas at San Antonio, 1976.

Hood, Charles, and James Underwood, Jr. "Geology of Palo Duro Canyon." In *The Story of Palo Duro Canyon,* ed. Duane Guy, 3–34. Canyon: Panhandle-Plains Historical Society, 1979.

Hughes, Jack T. "Archeology of Palo Duro Canyon." In *The Story of Palo Duro Canyon,* 35–58.

Hughes, Jack T., Gerald Etchieson, and R. Speer. *A Final Report of Archeological Survey . . . in Caprock Canyons State Park.* Austin: Texas Department of Parks and Wildlife, 1977.

Hughes, Jack T., and Eddie Guffee. *An Archeological Survey in the Running Water Draw Watershed.* Washington, D.C.: Report to the Soil Conservation Service, 1974.

Hughes, Jack T., and Patrick Willey. *Archeology at Mackenzie Reservoir.* Austin: Texas Historical Commission, 1978.

Johnson, Eileen, ed. *Lubbock Lake: Late Quaternary Studies on the Southern High Plains.* College Station: Texas A&M University Press, 1987.

Katz, Sussanna R., and Paul Katz. *Archeological Investigations in Lower Tule Canyon.* Archeology Survey Report, no. 16. Austin: Texas Historical Commission, 1976.

Krutch, Joseph Wood. *The Desert Year.* New York: William Sloane and Associates, 1951.

McDonald, Jerry. *North American Bison: Their Classification and Evolution.* Berkeley: University of California Press, 1981.

———. "The Reordered North American Selection Regime and Late Quaternary Megafaunal Extinctions." In *Quaternary Extinctions,* 404–439.

McFee, John. *Basin and Range.* New York: Farrar, Straus, Giroux, 1980.

———. *Rising From the Plains.* New York: Farrar, Straus, Giroux, 1986.

Martin, Paul. "Prehistoric Overkill: The Global Model." In *Quaternary Extinctions,* 354–403.

Martin, Paul, and Henry Wright, Jr., eds. *Pleistocene Extinctions: The Search for a Cause.* New Haven: Yale University Press, 1967.

Matthews, William A. *The Geologic Story of Palo Duro Canyon.* Bureau of Economic Geology Guidebook 8. Austin: University of Texas, 1983.

Maxwell, Ross A. *Geologic and Historic Guide to the State Parks of Texas.* Austin: Bureau of Economic Geology, University of Texas, 1970.

"Patriarch of the Aviary." *Time,* August 25, 1986.

Reeves, C. C., Jr. "Pleistocene Climate of the Llano Estacado." *Geological Notes,* 1965, 180–189.

Runte, Alfred. *National Parks: The American Experience.* 2d ed. Lincoln: University of Nebraska Press, 1987.

Schultz, Gerald. "The Paleontology of Palo Duro Canyon." In *The Story of Palo Duro Canyon,* 59–86.

Thomas, Ronny. *Geomorphic Evolution of the Pecos River System.* Baylor Geologic Studies Bulletin 22. Waco, 1972.

Turner, Frederick. *Beyond Geography: The Western Spirit against the Wilderness.* New York: Viking, 1980.
Wendorf, Fred, and James Hester, eds. *Late Pleistocene Environments of the Southern High Plains.* Rancho de Taos: Fort Burgoin Research Center, 1975.

2. SONG OF THE DESERT

Bilbo, Michael. "Return to Cowhead Mesa: The Proposed Cowhead Mesa National Register Archeological District." In *Transactions of the 21st Regional Archeological Symposium for Southeastern New Mexico and Western Texas,* ed. Carol Hendrick, 1–24. El Paso: El Paso Archaeological Society, 1986.
Box, Thadis. "Range Deterioration in West Texas." *Southwestern Historical Quarterly* 71 (1967): 39–45.
Boyd, Doug. "Archeology Survey of the Proposed Justiceburg Reservoir." Paper presented at the meeting of the South Plains Archeological Society, February 5, 1989.
Brown, Jimmy, and Joseph Schuster. "Effects of Grazing on a Hardland Site in the Southern High Plains." *Journal of Range Management* 22 (1969): 418–423.
Bryant, Fred, and Bruce Morrison. "Managing Plains Mule Deer in Texas and Eastern New Mexico." College of Agricultural Science Management Note 7. Lubbock: Texas Tech University, 1985.
Didway, Charles, ed. *Wagon Wheels: A History of Garza County.* Seagraves, Tex.: Pioneer Book Publishing, 1973.
Eaves, Charles Dudley. "Charles William Post, the Rainmaker." *Southwestern Historical Quarterly* 43 (1940): 425–437.
———. "Colonization Activities of Charles William Post." *Southwestern Historical Quarterly* 43 (1939): 72–84.
Haley, J. Evetts. *Charles Goodnight, Cowman and Plainsman.* Norman: University of Oklahoma Press, 1949.
Hickerson, Nancy. "The Jumano and Trade in the Arid Southwest, 1580–1700." Typescript in possession of the author.
Holden, William C. *Rollie Burns.* Dallas: Southwest Press, 1932.
Johnson, David. "Desert Buttes: Natural Experiments for Testing Theories of Island Biogeography." *National Geographic Research* 2 (Spring 1986): 152–166.
Jones, J. Knox, et al. *Annotated Checklist of Recent Land Mammals of Texas. Occasional Papers, no. 119.* Lubbock: The Museum, Texas Tech University Museum, 1988.
Kopp, April. "New Mexico's 'King of the Road.'" *New Mexico Magazine,* October 1987, 31–34.
Lawrence, D. H. *The Complete Poems of D. H. Lawrence.* Edited by Vivian de Sola Pinto and F. Warren Roberts. New York: Penguin Books, 1977.
LeFors, Rufe. *"Facts as I Remember Them:" The Autobiography of Rufe LeFors.* Edited by John Allen Peterson. Austin: University of Texas Press, 1986.
Mooar, J. Wright. File, Southwest Collection, Texas Tech University, Lubbock.
Quinn, Jean, and Jane Holden. "Caves and Shelters in Dawson and Borden Counties." *Texas Archeological and Paleontological Society Bulletin* 20 (1949): 114–131.
Riggs, Aaron. "Petroglyphs of Garza County, Texas." In *Transactions of the First Annual Archeological Symposium for Southeastern New Mexico and West Texas,* 9–14. Hobbs: Lea County Archeological Society, 1965.

———. "Yellowhouse Crossing Mesa Petroglyphs." In *Transactions of the Fifth Annual Regional Archeological Symposium for Southeastern New Mexico and West Texas,* 25–33. Portales: El Llano Archeological Society, 1969.

Shreve, Forrest. *The Plant Life of the Sonoran Desert.* Washington, D.C.: Carnegie Institute of Washington, 1936.

Stegner, Wallace. *The American West as Living Space.* Ann Arbor: University of Michigan Press, 1987.

Strickon, Arnold. "The Euro-American Ranching Complex." In *Man, Culture, and Animals,* ed. Anthony Leeds and Andrew Vayda, 229–258. Washington, D.C.: AAAS, 1965.

Tuan, Yi Fu. "Topophilia: Personal Encounters with the Landscape." In *Man, Space, and Environment,* ed. P. English and R. Mayfield, 538–558. New York: Oxford University Press, 1961.

Udall, Stewart. *To the Inland Empire.* New York: Doubleday, 1987.

Upshaw, Emily. "Palo Duro Rock Art: Indian Petroglyphs and Pictographs." M.A. thesis, West Texas State University, Canyon, 1972.

Vogt, Evon, and John Roberts. "A Study of Values." *Scientific American* 195 (July 1956): 25–30.

Weniger, Del. *Cacti of Texas and Neighboring States.* Austin: University of Texas Press, 1987.

Wiggers, Ernie, and Samuel Beason. "Characterization of Sympatric or Adjacent Habitats of 2 Deer Species in West Texas." *Journal of Wildlife Management* 50 (1986): 129–134.

Young, Stanley, and Edward Goldman. *The Puma.* New York: Dover, 1946.

3. GRASSY GORGES OF THE BRAZOS

Abbey, Edward. *Desert Solitaire: A Season in the Wilderness.* New York: Simon and Schuster, 1968.

Amangual, Francisco. "[Journal,] San Antonio to Santa Fe to San Elzeario to San Antonio, in 1808." In *Pedro Vial and the Roads to Santa Fe,* by Noel Loomis and Abraham Nasatir, 459–534. Norman: University of Oklahoma Press, 1967.

Bretting, P. K. "Changes in Fruit Shape in *Proboscidea parviflora ssp. parviflora* (Martyniaceae) with Domestication." *Economic Botany* 40 (1986): 170–176.

Brown, David, ed. *The Wolf in the Southwest: The Making of an Endangered Species.* Tucson: University of Arizona Press, 1983.

Carlson, Gustav G., and Volney H. Jones. "Some Notes on the Uses of Plants by the Comanche Indians." *Papers of the Michigan Academy of Sciences, Arts, and Letters* 25 (1940): 517–542.

Cook, John R. *The Border and the Buffalo.* Topeka: Crane and Co., 1907.

Curtin, Leonora Scott Muse. *Healing Herbs of the Upper Rio Grande.* Santa Fe Laboratory of Anthropology, 1947. Reprint. New York: Arno Press, 1976.

Dubos, René. *The Wooing of Earth.* New York: Charles Scribner's Sons, 1980.

DeBuys, William. *Enchantment and Exploitation: The Life and Hard Times of a New Mexico Mountain Range.* Albuquerque: University of New Mexico Press, 1985.

Favour, Aldepheus. *Old Bill Williams: Mountain Man.* Norman: University of Oklahoma Press, 1962.

Fenton, James. "The Lobo Wolf: Beast of Waste and Desolation." *Panhandle-Plains Historical Review* 53 (1980): 57–70.

Haley, J. Evetts, ed. *Albert Pike's Journeys in the Prairie, 1831–1832.* Canyon: Panhandle-Plains Historical Society, 1969.

———. "Grass Fires of the Southern Plains." *West Texas Historical Association Yearbook* 5 (1929): 18–34.

Haynes, Mike. "Account Told of Miles-Long Wolf Pack." *Dallas Morning News*, March 30, 1986.

Harrell Family Papers. "The Last Wolf Killed in the Panhandle." Oral History Tape, Panhandle-Plains Historical Museum Archives, Canyon.

Hesse, Hermann. *Steppenwolf*. 1929. Reprint. New York: Holt, Rinehart & Winston, 1963.

Jones, David. *Sanapia: Comanche Medicine Woman*. New York: Holt, Rinehart & Winston, 1972.

Kavanagh, Thomas. "Political Power and Political Organization: Comanche Politics, 1786–1875." Ph.D. dissertation, University of New Mexico, Albuquerque, 1986.

Kerouac, Jack. *Mexico City Blues*. New York: Grove Press, 1960.

Lewis, Henry. "Why Indians Burned: Specific versus General Reasons." In *Proceedings of a Symposium and Workshop on Wilderness Fire*, 75–80. Ogden: U.S. Forest Service General Technical Report INT-182, 1985.

Mares, José. "[Journal,] Santa Fe to Bexar, July 31 to October 8, 1787, and Return, in 1788." In *Pedro Vial and the Roads to Santa Fe*, 288–315.

Momaday, N. Scott. "The Man Made of Words." In *The Remembered Earth*, ed. Geary Hobson, 162–173. Albuquerque: University of New Mexico Press, 1981.

Mowrey, Daniel, and Paul Carlson. "The Native Grasslands of the High Plains of West Texas: Past, Present, Future." *West Texas Historical Association Yearbook* 63 (1987): 24–41.

Nathan, Gary. "Note on Galactomannan." *Outside*, November 1986, 48.

Nichols, John. *The Last Beautiful Days of Autumn*. New York: Holt, Rinehart & Winston, 1982.

Pyne, Stephen. *Fire in America: A Cultural History of Wildland and Rural Fire*. Princeton: Princeton University Press, 1982.

Rathjen, Fred. *The Texas Panhandle Frontier*. Austin: University of Texas Press, 1973.

Roosevelt, Theodore. "Buffalo Hunting." *St. Nicholas* 17 (December 1889): 136–143.

Rose, Francis, and Russell Strandtmann. *Wildflowers of the Llano Estacado*. Dallas: Taylor Publishing Co., 1986.

Snyder, Gary. "Reinhabitation." In *The Old Ways*, 57–66. San Francisco: City Lights Books, 1977.

Strickland, Rex, ed. "The Recollections of W. S. Glenn, Buffalo Hunter." *Panhandle-Plains Historical Review* 22 (1949): 15–62.

Teale, Edwin Way. *Journey into Summer*. New York: Dodd, Mead, 1960.

Thoreau, Henry David. *Walden, or Life in the Woods*. 1854. New York: Anchor Books, 1973.

———. *A Week on the Concord and Merrimack Rivers*. Vol. 1 of *The Writings of Henry David Thoreau*. Boston: Riverside Press, 1893.

Vestal, Paul A., and Richard Schultes. *The Economic Botany of the Kiowa Indians*. Cambridge, Mass.: Botanical Museum of Harvard University, 1939.

Wallace, Ernest, ed. "The Journal of Ranald S. Mackenzie's Messenger to the Kwahadi Comanches." *Red River Valley Historical Review* 3 (1978): 227–246.

Waters, John. "The Synoptic Climatology of the South Plains Dryline over the Period April–June, 1970–1979." M.S. thesis, Texas Tech University, Lubbock, 1987.

Weaver, J. E., and F. W. Albertson. *Grasslands of the Great Plains*. Lincoln: Johnsen Publishing Co., 1956.

Whitlock, V. H. *Cowboy Life on the Llano Estacado.* Norman: University of Oklahoma Press, 1970.
Wright, Henry, and Arthur Bailey. *Fire Ecology.* New York: John Wiley & Sons, 1982.
Yellowhouse Canyon. File, Southwest Collection, Texas Tech University, Lubbock.
Young, Stanley, and E. A. Goldman. *The Wolves of North America.* 2 vols. New York: Dover, 1944.

4. *LOS CAÑONES DEL VALLE DE LAS LÁGRIMAS*
THE CANYONS OF THE VALLEY OF TEARS

Austin, Mary. *The American Rhythm.* New York: Houghton Mifflin, 1923.
Barsness, Larry. *Heads, Hides, and Horns: The Compleat Buffalo Book.* Fort Worth: Texas Christian University Press, 1974.
Basso, Keith. "'Stalking with Stories': Names, Places, and Moral Narratives among the Western Apaches." In *On Nature,* ed. Daniel Halpern. San Francisco: North Point Press, 1987.
Berlandier, Jean Louis. *The Indians of Texas in 1830.* Edited by John C. Ewers. Washington, D.C.: Smithsonian Institution Press, 1969.
Berger, Thomas. *Little Big Man.* New York: Dial Press, 1964.
Boyd, Maurice. *Kiowa Voices: Ceremonial Dance, Ritual and Songs.* 2 vols. Fort Worth: Texas Christian University Press, 1981.
Brown, William. "Comanchería Demography, 1805–1830." *Panhandle-Plains Historical Review* 59 (1986): 1–17.
Castaneda, Carlos. *Journey to Ixtlan: The Lessons of Don Juan.* New York: Simon and Schuster, 1972.
Chapman, Joseph, and George Feldhamer, eds. *Wild Animals of North America.* Baltimore: Johns Hopkins University Press, 1972.
Crawford, Max. *Lords of the Plain.* New York: Atheneum, 1985.
de Vaca, Cabeza. "The Narrative of Cabeza de Vaca." In *Spanish Explorers in the Southern United States, 1528–1543,* ed. F. W. Hodge, 1–126. Austin: Texas State Historical Association, 1984.
Dobie, J. Frank. *The Mustangs.* New York: Bramhall House, 1934.
Dobyns, Henry. *Their Number Become Thinned.* Knoxville: University of Tennessee Press, 1983.
Drinnon, Richard. "The Metaphysics of Dancing Tribes." In *The American Indian and the Problem of History,* ed. Calvin Martin, 106–114. New York: Oxford University Press, 1987.
Ewers, John C. "The Influence of Epidemics on the Populations and Cultures of Texas." *Plains Anthropologist* 18 (1973): 107–121.
Flores, Dan. "Comanche Ecology." Paper presented at the Symposium on the Llano Estacado Experience, Texas Tech University, 1986.
———. "Southern Plains Indian Ecology and Bison." Paper presented at the annual meeting of the Organization of American Historians, Reno, Nevada, 1988.
Fowler, Jacob. *The Journal of Jacob Fowler.* Edited by Elliott Coues. New York: Francis P. Harper, 1898.
Gelo, Daniel, and Melburn Thurman. "Critique and Response to Thurman's 'On a New Interpretation of Comanche Social Organization.'" *Current Anthropology* 28 (1987): 551–555.
Goss, James."Basin-Plateau Shoshonean Ecological Model." In *Great Basin Cultural Ecology: A Symposium,* ed. Don Fowler, 123–128. Reno: Desert Research Institute, 1972.

Guffee, Eddie. *The Merrill-Taylor Village Site: An Archeological Investigation of Pre-Anglo, Spanish-Mexican Occupation on Quitaque Creek in Floyd County, Texas.* Austin: Texas Historical Commission, 1976.

Halloran, Arthur. "Bison (Bovidae) Productivity on the Wichita Mountains Wildlife Refuge, Oklahoma." *Southwestern Naturalist* 13 (1968): 23–26.

Harrison, Lowell, ed. "Three Comancheros and a Trader." *Panhandle-Plains Historical Review* 38 (1965): 75–93.

Kaplan, David. "The Law of Cultural Dominance." In *Evolution and Culture,* ed. Marshall Sahlins and Elman Service, 67–81. Ann Arbor: University of Michigan Press, 1960.

Kardiner, Abraham. "Analysis of Comanche Culture." In *The Psychological Frontiers of Society,* by Abraham Kardiner, et al. New York: Columbia University Press, 1945.

Kavanagh, Thomas. "The Comanches: Paradigmatic Anomaly or Ethnographiuc Fiction?" *Haliksa'i* 4 (1985): 109–128.

Kelton, Elmer. *The Wolf and the Buffalo.* New York: Doubleday, 1980.

Kenner, Charles. *A History of New Mexican–Plains Indian Relations.* Norman: University of Oklahoma Press, 1969.

Krysl, L. J., et al. "Horses and Cattle Grazing in the Wyoming Red Desert: I. Food Habits and Dietary Overlap." *Journal of Range Management* 37 (1984): 72–76.

LaBarre, Weston. *The Peyote Cult.* 4th ed. New York: Schocken Books, 1975.

Levine, Francis, and Martha Freeman. *A Study of Documentary and Archeological Evidence for Comanchero Activity in the Texas Panhandle.* Austin: Texas Historical Commission, 1982.

Levy, Jerold. "Ecology of the South Plains." In *Symposium: Patterns of Land Use and Other Papers,* ed. Viola Garfield, 18–25. Seattle: University of Washington Press, 1961.

McHugh, Tom. *Time of the Buffalo.* New York: Alfred A. Knopf, 1972.

McMurty, Larry. *Lonesome Dove.* New York: Simon and Schuster, 1985.

Martin, Calvin. "An Introduction aboard the Fidele" and "Time and the American Indian." In *The American Indian and the Problem of History,* 3–26, 192–220.

Michener, James A. *Texas.* Austin: University of Texas Press, 1986.

Moore, John. *The Cheyenne Nation.* Lincoln: University of Nebraska Press, 1987.

Oliver, Symmes. *Ecology and Cultural Continuity as Contributing Factors in the Social Organization of the Plains Indians.* Berkeley: University of California Press, 1972.

Roe, Frank G. *The North American Buffalo: A Study of the Species in Its Wild State.* Toronto: University of Toronto Press, 1951.

Sessions, George. "The Deep Ecology Movement: A Review." *Environmental Ethics* 11 (1987): 105–125.

Shepard, Paul. *The Tender Carnivore and the Sacred Game.* New York: Scribner's, 1973.

Shimkin, Dimitri. *Ethnogeography of the Wind River Shoshone.* Berkeley: University of California Press, 1947.

Shull, Alisa, and Alan Tipton. "Effective Population Size of Bison on the Wichita Mountains Wildlife Refuge." *Conservation Biology* 1 (May 1987): 35–41.

Smith, Henry Nash. *Virgin Land: The American West in Symbol and Myth.* Cambridge, Mass." Harvard University Press, 1950.

Spykerman, Brian Ray. "Shoshone Conceptualizations of Plant Relationships." M.S. thesis, Utah State University, Logan, 1977.

St. Clair Robson, Lucia. *Ride the Wind.* New York: Ballentine Books, 1982.

Thompson, H. Paul. "A Technique Using Anthropological and Biological Data." *Current Anthropology* 7 (1966): 417–424.
Tolbert, Frank X. *The Staked Plain.* Reprint. College Station: Texas A&M University Press, 1987.
Toynbee, Arnold. "The Religious Background of the Present Environmental Crisis." In *Ecology and Religion in History,* ed. David and Eileen Spring, 137–150. New York: Harper & Row, 1974.
Twain, Mark. "The French and the Comanches." In *Complete Essays of Mark Twain.* Edited by Charles Nerden. Garden City, N.Y.: Doubleday, 1963.
Wallace, Ernest, and E. Adamson Hoebel. *The Comanches: Lords of the South Plains.* Norman: University of Oklahoma Press, 1952.
Wallace, Ernest, ed. *Ranald S. Mackenzie's Official Correspondence, 1871–1873.* Lubbock: West Texas Museum Association, 1967.
Webber, Charles. *Old Hicks the Guide.* Upper Saddle River, N.J.: Literature House, 1855.
White, Richard, guest ed. *Environmental Review 9. Special Issue: Native Americans and the Environment,* Summer 1985.
Zubrow, Ezra. *Prehistoric Carrying Capacity: A Model.* Menlo Park, Calif.: Cummings Publishing, 1975.

5. WILDERNESS CATHEDRALS

Abbey, Edward. "Freedom and Wilderness, Wilderness and Freedom." In *The Journey Home: Some Words in Defense of the American West,* 227–238. New York: E. P. Dutton, 1977.
Adams, David. "Vegetation-Environment Relationships in Palo Duro Canyon, West Texas." Ph.D. dissertation, University of Oklahoma, Norman, 1979.
Bailey, Vernon. *Biological Survey of Texas.* Washington, D.C.: Government Printing Office, 1905.
Baker, T. Lindsay, ed. "The Survey of the Headwaters of the Red River, 1876." *Panhandle-Plains Historical Review* 58 (1985): 1–124. Includes the official report of Lieutenant Ernest Ruffner, previously published in *Report of the Secretary of War,* vol. 2, pt. 2, in U.S. Congress House Exec. Doc. 1, pt. 2, 45th Cong. 2nd sess. (1877), and the previously unpublished diary of Carl J. A. Hunnius, topographical illustrator of the expedition.
Bolen, Eric, and Dan Flores. "The Mississippi Kite in The Environmental History of the Southern Great Plains." *Prairie Naturalist* 21 (1989): 65–74.
Brown, David. *The Grizzly in the Southwest.* Norman: University of Oklahoma Press, 1985.
Douglas, William O. *A Farewell to Texas: A Vanishing Wilderness.* New York: McGraw-Hill, 1967.
Estep, Raymond. "The Le Grand Survey of the High Plains—Fact or Fancy?" *New Mexico Historical Review* 29 (1954): 81–96. Includes Le Grand's journal in "Notes and Documents," pp. 141–153.
Flores, Dan, ed. *Journal of an Indian Trader: Anthony Glass and the Texas Trading Frontier, 1790–1810.* College Station: Texas A&M University Press, 1985.
Fritz, Edward, and Jess Alford. *Realms of Beauty: The Wilderness Areas of East Texas.* Austin: University of Texas Press, 1986.
Goetzmann, William. *Exploration and Empire: The Explorer and the Scientist in the Winning of the American West.* New York: W. W. Norton, 1966.
Harris, David. "Recent Plant Invasions in the Arid and Semi-arid Southwest of the United States." In *Man's Impact on Environment,* ed. Thomas Detwyler, 459–481. New York: McGraw-Hill, 1971.

Hollon, Eugene. *Beyond the Cross Timbers: Travels of Randolph B. Marcy, 1812–1887.* Norman: University of Oklahoma Press, 1955.

McCauley, C. A. H. "Notes on the Ornithology of the Region about the Source of the Red River of Texas." Edited by Kenneth Seyffert and T. Lindsay Baker. *Panhandle-Plains Historical Review* 61 (1988): 25–88. McCauley's account, edited by Elliott Coues, first appeared in Ferdinand V. Hayden, *Bulletin of the United States Geological and Geographical Survey of the Territories,* III, no. 2. Washington, D.C.: Government Printing Office, 1877.

Marcy, Randolph. *A Report on the Exploration of the Red River, in Louisiana.* Washington, D.C.: Government Printing Office, 1854.

Mooar, J. Wright. "Frontier Experiences of J. Wright Mooar." *West Texas Historical Association Yearbook* 4 (1928): 89–92.

Nash, Roderick. *Wilderness and the American Mind.* 3d ed. New Haven: Yale University Press, 1982.

Peterson, Roger Tory. *A Field Guide to the Birds of Texas and Adjacent States.* Boston: Houghton Mifflin, 1960.

Seyffert, Kenneth, et al. *A Checklist of the Birds of Palo Duro Canyon State Park.* Austin: Texas Parks and Wildlife Department, 1975.

———. *Birds of Caprock Canyons State Park.* Austin: Texas Parks and Wildlife Department, 1982.

Simpson, C. David, ed. *Proceedings of the Symposium on Ecology and Management of Barbary Sheep.* Lubbock: Texas Tech Press, 1980.

True, Dan. *A Family of Eagles.* New York: Everest House, 1980.

Whipple, Thomas King. *Study Out the Land.* Berkeley: University of California Press, 1943.

Wright, Robert. "The Vegetation of Palo Duro Canyon." In *The Story of Palo Duro Canyon,* 87–116.

6. VISIONS OF PALO DURO AND OTHER CANYONS OF THE IMAGINATION

Castro, Jan. *The Art and Life of Georgia O'Keeffe.* New York: Crown Publishers, 1985.

Cowart, Jack, Juan Hamilton, and Sarah Greenough. *Georgia O'Keeffe: Art and Letters.* Washington, D.C.: National Gallery of Art, 1987.

DeLong, Lea. *Nature's Forms, Nature's Forces: The Art of Alexandre Hogue.* Norman: University of Oklahoma Press and the Philbrook Museum, 1984.

Dutton, Clarence. *Tertiary History of the Grand Canyon District.* Washington, D.C.: U.S. Geological Survey Report, 1882.

Flores, Dan. "Canyons of the Imagination." *Southwest Art* 18 (March 1989): 70–76.

Gibson, Arrell. *The Santa Fe and Taos Colonies: Age of the Muses, 1900–1924.* Norman: University of Oklahoma Press, 1983.

Goetzmann, William H. *William H. Holmes Panoramic Art.* Fort Worth: Amon Carter Museum, 1977.

Goetzmann, William H., and William N. Goetzmann. *The West of the Imagination.* New York: W. W. Norton, 1986.

Haley, James. *The Buffalo War.* Norman: University of Oklahoma Press, 1985.

Hartman, Bruce, and Susie Kalil. *Frank Reaugh: The Southwestern Landscape.* Canyon: Panhandle-Plains Historical Museum, 1986.

Hogue, Alexandre. "Palo Duro: The Paradise of the Panhandle." *Dallas Times-Herald,* July 24, 1927.

Hunnius, Carl J. A. "Survey of the Sources of the Red River, April 25th to June 30th 1876. In "The Survey of the Headwaters of the Red River, 1876." *Panhandle-Plains Historical Review* 58 (1985): 52–121.

Jackson, John Brinckerhoff. *The Southern Landscape Tradition in Texas.* Fort Worth: Amon Carter Museum, 1980.
Kalil, Susie. *The Texas Landscape, 1900–1986.* Houston: Museum of Fine Arts, 1986.
Kandinsky, Wassily. *Complete Writings on Art.* Edited by Kenneth Lindsay and Peter Vergo. Boston: G. K. Hall, 1982.
Kendall, George Wilkins. *Narrative of the Texan-Santa Fe Expedition.* 2 vols. 1842. Facsimile ed. Austin: Steck Co., 1935.
Lisle, Laurie. *Portrait of an Artist: A Biography of Georgia O'Keeffe.* New York: Washington Square Press, 1980.
Lowes, Ruth, and W. Mitchell Jones. *We'll Remember Thee: An Informal History of West Texas State University.* Canyon: Panhandle-Plains Historical Museum, 1984.
Mead, Ben Carleton. File, Panhandle-Plains Historical Museum Archives, Canyon.
Mills, Enos. "Address to Amarillo Kiwanis Club on Palo Duro as a Park." *Amarillo Daily News,* October 20, 1921.
Muir, John. *The Mountains of California.* Rev. ed. New York: Century Co., 1911.
Nelson, Mary Carroll. "Daryl Howard's Landscape Woodcuts." *American Artist,* December 1986, 58–63, 102.
Novak, Barbara. *Nature and Culture: American Landscape and Painting, 1825–1875.* New York: Oxford University Press, 1979.
O'Keeffe, Georgia. *Georgia O'Keeffe.* New York: Viking 1976.
Reaugh, Frank. Papers. Panhandle-Plains Historical Museum Archives, Canyon.
Shepard, Paul. *Man in the Landscape.* New York: Alfred A. Knopf, 1967.
Stewart, Rick. *Lone Star Regionalism: The Dallas Nine and Their Circle.* Austin: Texas Monthly Press, 1985.
Udall, Sharon Rohlfson. *Modernist Painting in New Mexico, 1913–1985.* Albuquerque: University of New Mexico Press, 1985.
Warner, Phebe. "Palo Duro—as a National Park." *Southwest Plainsman,* November 16, 1929.
———. "The Mission of Our Palo Duro Canyon." *Southwest Plainsman,* November 21, 1930.
Wilson Hurley: A Retrospective Exhibition. Kansas City, Mo.: Lowell Press, 1985.

7. DUST-BLOWN DREAMS AND THE CANADIAN RIVER GORGE

Bowden, Charles. *Killing the Hidden Waters.* Austin: University of Texas Press, 1977.
Cabeza de Baca, Fabiola. *We Fed Them Cactus.* Albuquerque: University of New Mexico Press, 1954.
Calvin, William. *The River That Flows Uphill.* San Francisco: Sierra Club Books, 1986.
Carroll, H. Bailey, ed. *The Journal of Lieutenant J. W. Abert from Bent's Fort to St. Louis in 1845.* Canyon: Panhandle-Plains Historical Museum, 1941.
Dostoyevsky, Fyodor. *The Brothers Karamozov.* Translated by Constance Garnett. New York: Signet Books, 1957.
Fincher, Jack. "A Home Where the Equids Can Roam." *Smithsonian* 18 (May 1987): 138–153.
Glanz, Michael, and Jesse Ausubel. "The Ogallala Aquifer and Carbon Dioxide: Comparison and Convergence." *Environmental Conservation* 2 (Summer 1984): 123–131.

Governors Papers. Tingley File. New Deal Agencies Files, New Mexico State Archives, Santa Fe, New Mexico.
Green, Donald. *Land of the Underground Rain: Irrigation on the Texas High Plains, 1910–1970.* Austin: University of Texas Press, 1973.
High Plains Associates. *Six State High Plains Ogallala Aquifer Regional Resources Study.* Austin: HPA, 1982.
High Plains Underground Water Conservation Districts. *Hydrologic Atlases,* 1980–1988.
Kromm, David, and Stephen White. *Conserving the Ogallala: What Next?* Manhattan: Kansas State University, 1985.
New Mexico State Engineer Files, New Mexico State Archives, Santa Fe, New Mexico.
Momaday, N. Scott. *The Way to Rainy Mountain.* Albuquerque: University of New Mexico Press, 1969.
Pearce, T. M., ed., *New Mexico Place Names.* Albuquerque: University of New Mexico Press, 1965.
Roper, Beryl. "How Did This River Come to Be Called Canadian." *Panhandle-Plains Historical Review* 61 (1988): 17–24.
Tidwell, Dewey. "Nearly a Ghost." *Grain Procedures News* 31 (1980): 13–15.
Webb, Walter Prescott. "The American West: Perpetual Mirage." *Harper's Magazine* 214 (1957): 25–31.
———. *The Great Plains: A Study in Institutions and Environment.* Boston: Dodd and Mead, 1931.
Works Progress Administration Files, County Histories (Harding and Mora), Rivers and Canyons, New Mexico State Archives, Santa Fe, New Mexico.
Worster, Donald. *Dust Bowl: The Southern Plains in the 1930s.* New York: Oxford University Press, 1979.

8. THE MYTHIC, THE SACRED, THE PRESERVABLE

Attfield, Robin. "Christian Attitudes to Nature." *Journal of the History of Ideas* 44 (1983): 369–386.
Barr, James. "Man and Nature: The Ecological Controversy and the Old Testament." In *Ecology and Religion in History,* 15–31.
Berkhofer, Robert. "Space, Time, Culture, and the New Frontier." *Agricultural History* 38 (January 1964): 21–30.
Campbell, Joseph, and Bill Moyers. *The Power of Myth.* New York: Doubleday, 1988.
Carls, Glenn E., and A. K. Ludeke. "State Park Contributions to Natural Areas Protection in Texas, USA." *Biological Conservation* 28 (1984): 95–110.
Cotten, Don. "Growing Grapes on the South Plains." *Lubbock Lights* 1 (August 1988): 6–9.
Deknatel, Charles. "Regionalism and Environment: The Search for Planning Strategy and Organization in the Great Plains." *Environmental Review* 10 (1986): 107–121.
Drinnon, Richard. *Facing West: The Metaphysics of Indian-Hating and Empire Building.* Minneapolis: University of Minnesota Press, 1980.
Evernden, Neil. "Beauty and Nothingness: Prairie as Failed Resource." *Landscape* 27 (1983): 3–8.
Flores, Dan, and Blake Morris. "The Plains and the Parks: Palo Duro Canyon and a Failed Attempt at a 'National Park of the Plains.'" In *The National Parks: A Historical Reader,* ed. Stephen Mehls and Michael Schene. Lawrence: University of Kansas Press, forthcoming.

Hughes, J. Donald, and Jim Swan. "How Much of the Earth Is Sacred Space?" *Environmental Review* 10 (Winter 1986): 247–259.
Jameson, John. *Big Bend on the Rio Grande: Biography of a National Park.* New York: Peter Lang Publishing, 1987.
———. "The Quest for a National Park in Texas." *West Texas Historical Association Yearbook* 50 (1974): 47–60.
Limerick, Patricia. *The Legacy of Conquest: The Unbroken Past of the American West.* New York: W. W. Norton, 1987.
Mead, James. *Hunting and Trading on the Great Plains, 1859–1875.* Edited by Schuyler Jones. Norman: University of Oklahoma Press, 1986.
Miller, Kenton. "The Natural Protected Areas of the World." In *National Parks, Conservation, and Development: The Role of Protected Areas in Sustaining Society,* ed. Jeffry McNeely and Kenton Miller, 17–22. Washington, D.C.: Smithsonian Institution Press, 1984.
National Park Service. Papers, National Archives Record Group 79, Washington, D.C.
Newman, John. "Llano Estacado Heritage: Towards a Regional Cultural-Development Interpretive System." M.A. thesis, Texas Tech University, Lubbock, 1974.
Petersen, Peter. "A Park for the Panhandle." In *The Story of Palo Duro Canyon,* 145–178.
Popper, Deborah Epstein, and Frank Popper. "A Daring Proposal for Dealing with an Inevitable Disaster." *Planning,* December 1987, 12–18.
———. "The Fate of the Plains." *High Country News* 20 (September 26, 1988): 15–19.
Reisner, Marc. *Cadillac Desert.* New York: Viking, 1986.
Sale, Kirkpatrick. "Deep Ecology and Its Critics." *The Nation,* May 14, 1988, 27–30.
Sessions, George. "Deep Ecology and the New Age." *Earth First!,* September 23, 1987, 27–30.
Six Year Plan, 1988–1993. Austin: Texas Parks and Wildlife Department, 1988.
Slotkin, Richard. *Regeneration through Violence: The Mythology of the American Frontier, 1600–1860.* Middleton, Conn.: Wesleyan University Press, 1973.
Toll, Roger. "Investigative Report on Proposed Palo Duro National Park." National Park Service Papers, NARG 79, File 0-32, Box 2947. Washington, D.C.
Tolson, Hillory, Milo Christianson, et al. "Investigative Report on Proposed Palo Duro National Monument Texas." Washington, D.C.: National Park Service Report, 1939.
Wallach, Bret. "The Return of the Prairie." *Landscape* 28 (1985): 1–5.
White, Lynn, Jr. "The Historical Roots of Our Ecologic Crisis." In *Ecology and Religion in History,* 15–31.
Whitman, Walt. *Leaves of Grass.* 1855. New York: Bantam Books, 1983.
Williams, Dennis. "John Muir, Christian Mysticism, and the Spiritual Value of Nature: 1866–1873." M.A. thesis, Texas Tech University, Lubbock, 1989.

Index